D1336756

TO BE
DISPOSED
BY
AUTHORITY

STUDIES IN THEORETICAL AND APPLIED ECONOMICS
General editor:
B. T. BAYLISS

THE BRITISH POTTERY INDUSTRY

P. W. GAY AND R. L. SMYTH
Lecturer and Senior Lecturer in Economics, University of Keele

LONDON
BUTTERWORTHS

THE BUTTERWORTH GROUP

ENGLAND
Butterworth & Co (Publishers) Ltd
London: 88 Kingsway, WC2B 6AB

AUSTRALIA
Butterworths Pty Ltd
Sydney: 586 Pacific Highway, NSW 2067
Melbourne: 343 Little Collins Street, 3000
Brisbane: 240 Queen Street, 4000

CANADA
Butterworth & Co (Canada) Ltd
Toronto: 14 Curity Avenue, 374

NEW ZEALAND
Butterworths of New Zealand Ltd
Wellington: 26–28 Waring Taylor Street, 1

SOUTH AFRICA
Butterworth & Co (South Africa) (Pty) Ltd
Durban: 152–154 Gale Street

First published 1974

© Butterworth & Co (Publishers) Ltd, 1974

ISBN 0 408 70565 5

Typeset by Amos Typesetters, Hockley, Essex
Printed and bound in England by
Butler & Tanner Ltd, Frome and London

FOREWORD

Pottery, the section of ceramic production which produces tableware, ceramic wall and floor tiles, sanitary ware and porcelain insulators, has evolved from a craft-based industry into a modern complex of economic and technical significance. Mr. Smyth and Mr. Gay have made the first serious attempt to analyse this evolution and to quantify its activities to date. The result is a fascinating record in breadth and depth and one which will serve as an important reference for those who wish to study an industry which contributes so much, in both economic and aesthetic terms, to our standards and way of life.

This book should not be received as the finite on the subject, but it will stand as helpful and interesting. I believe this book to be unique as an economic study of one of the world's modern industries.

S. H. Jerrett, O.B.E., T.D., D.L., C.I.Ceram.
Director, British Ceramic Manufacturers' Federation

PREFACE

There are numerous books on collecting porcelain and bone china and on pottery firms in the eighteenth and nineteenth centuries, but not on how the pottery industry is organised and operates here and now. There are also numerous books by economists on British industries, from coal and cotton to chemicals and electronics, but not one on pottery. This is a book that had to be written, and we are pleased to have had the opportunity of doing so.

Research was undertaken into economic aspects of the neighbouring pottery industry by Professor Bruce Williams, the first Professor of Economics at the University of Keele (University College of North Staffordshire, 1949–1962). His account of the industry was published in a symposium edited by Duncan Burn, *The Structure of British Industry*, and published by the Cambridge University Press in 1958. Between 1967 and 1971 I was involved in research projects concerned with exports, structure, home demand, workers and wages, and job satisfaction in the pottery industry. Five research reports on the industry were published by the Department of Economics, University of Keele. In this book we draw on material contained in the reports. We have not, however, been content merely to reproduce material from them; we have attempted to break new ground and criticism of the reports was important in persuading us to dispense with some ideas and modify others. Even when old ideas and propositions persist, at least they have been expressed, we hope, more precisely and perhaps rather more elegantly.

We are most grateful to the Social Science Research Council for their research grants, which enabled the project to be started and to continue for four years. We are also grateful to the University of Keele for providing additional finance to enable the research programme to be completed.

Mr. Gay and myself are most grateful to the research assistants who worked with us (or, as they would say, 'did the work')—Valerie Irvine, Christine Lilleker (now Christine Edwards), Hugh Macdiarmid and Dennis Gregory. We deeply regret the death of David Price in 1968. A native of Stoke-on-Trent and a graduate of the University of Leeds, he had worked with us for only three weeks as a research assistant when he was drowned.

Mr. D. J. Machin, who is a Senior Lecturer in Ceramics at the North Staffordshire Polytechnic, worked closely with us on all aspects of technology and innovation. He continues to investigate blungers, grinders, dryers, etc. with great gusto and insight, much to our benefit. We are most grateful for Mr. Machin's generosity; he has raised no objections to our presenting some of his ideas as if they were our own.

We have been sustained throughout our investigations by Mr. Sam Jerrett, the Director of the British Ceramic Manufacturers' Federation, and we are most grateful for his Foreword to this book. Mr. E. A. Batchelor, the Director of the Council of British Sanitaryware Manufacturers, also helped us. Reluctantly we have refrained from quoting from his surveys of cistern noise and loo provision. Again and again pottery managers have generously given us their undivided attention to explain why a particular process has not yet been mechanised or how new designs are introduced. Much of what we have written about the industry is based on what potters have told us and shown us. We are most grateful for the willingness of businessmen to talk at length with academic economists. Unfortunately, only some of their firms are adequately featured in the text. We are also grateful to the Secretary of the Ceramic and Allied Trades Union, Mr. A. Dulson, for his interest and advice.

Throughout the period during which we have worked on the industry we have received considerable assistance from the staff of the University Computer Centre, and much advice and encouragement from Professors E. M. Hugh-Jones and L. Fishman of the Economics Department. Finally, we are most grateful to Miss Joy Cooke and Miss Valerie Wilkinson for their assistance in typing the manuscript.

University of Keele   *R. L. Smyth*

ACKNOWLEDGEMENTS

We are grateful to the following publishers for permission to quote from the following books:
Aldus Books Ltd., *Ceramics in the Modern World*, by Maurice Chandler. (This is a particularly attractive and useful paperback on all aspects of ceramic technology.)
Basil Blackwell, *The Theory of the Growth of the Firm*, by Edith Penrose; George Allen & Unwin Ltd., *Export Performance and the Pressure of Demand*, by R. A. Cooper, K. Hartley and C. R. M. Harvey; Gerald Duckworth & Co. Ltd., *The Competitive Process*, by Jack Downie; Andre Deutsch Ltd., *The New Industrial State*, by J. K. Galbraith; and Macmillan & Co. Ltd. for permission to quote from *Manufacturing Business*, by P. W. S. Andrews and *Problems of a Mature Economy*, by F. V. Meyer, D. C. Corner and J. E. S. Parker.

We are grateful to Desmond Eyles for permission to quote from *Royal Doulton 1815–1965* which was published by Hutchinson of London in 1965.

We are also grateful to the London School of Economics for permission to quote from the article 'The Pricing of Manufactured Products' by Ronald S. Edwards, which appeared in *Economica* in August 1952, and to the Scottish Economic Society for permission to quote from my article 'Theories of Competition and the British Pottery Industry', which appeared in the *Scottish Journal of Political Economy* in February 1971.

The Central Statistical Office gave permission for us to reproduce *Figures* 9.2 and 9.3 which were published in *Economic Trends*.

We have found *Tableware International* (formerly *Pottery Gazette and Glass Trade Review*) invaluable in keeping us in touch with the latest developments in design and wholesaling and retailing, and we are grateful for the Editor's permission to quote extensively from an article on selling pottery in the United States. We are grateful to the Editor of *The Observer* to publish an extract from his newspaper.

Photographs have been included which illustrate various aspects of production and distribution of domestic pottery and industrial ceramics. We are grateful to the following firms for providing photographs:

Allied English Potteries Ltd.
Denbyware Ltd.
Doulton & Co. Ltd.
H. R. Johnson-Richards Tiles Ltd.
Staffordshire Potteries Ltd.
Twyfords Holdings Ltd.
Wedgwood Ltd.

Finally, but by no means least, we wish to thank Moodies Services Ltd. for giving us permission to quote from their summaries of company accounts.

CONTENTS

PART I COMPETITION, TECHNOLOGY AND   1
          ORGANISATION

1 THE STRUCTURE OF THE POTTERY INDUSTRY   3

          Output and Employment   4
          Exports   10
          Factory Size   11
          External Economies   12
          The Potteries   13
          Competition   15
          Summary   16

2 TECHNOLOGY   18

          Materials   19
          Manufacture   21
          Specialisation   34

3 COMPETITION   36

          Family Businesses   36
          Concentration   41
          Wedgwood Ltd.   43

Doulton & Co. Ltd. 44

Allied English Potteries Ltd. 45

Takeovers and Amalgamations 46

Management Teams 49

Competition and Monopoly 50

Trade and Research Associations 51

PART II THE DOMESTIC SECTOR OF THE INDUSTRY 53

4 THE 'BIG THREE' 55

Wedgwood Ltd. 59

Doulton & Co. Ltd. 69

Allied English Potteries Ltd. 74

The Royal Worcester Porcelain Company Ltd. 80

5 DOMESTIC WARE MANUFACTURERS 84

Spode Ltd. 85

Minton Ltd. 88

Mason's Ironstone China Ltd. 88

Staffordshire Potteries (Holdings) Ltd. 90

Alfred Clough Ltd. 95

J. & G. Meakin Ltd. 100

T. G. Green Ltd. 105

Thomas Poole & Gladstone China Ltd. 106

Arthur Wood & Son (Longport) Ltd. 109

Elijah Cotton Ltd. 111

Portmerion Potteries Ltd. 112

Denbyware Ltd. 113

6 COSTS 118

The Structure of Costs 119

Import Content 125

Management and Costs 126

The 'Seconds' Problem 130

7 THE HOME MARKET 132

Distribution 132

Expenditure on Tableware 135

Promotion		145
Pricing		146

8 EXPORTS 148

Organisation	156
Export Performance and the Pressure of Demand	161
Entry into Europe	162
The Dollar	163

PART III   TILES, SANITARY WARE, ELECTRICAL   165
             WARE AND INDUSTRIAL CERAMICS

9 INDUSTRIAL CERAMICS 167

Tiles	167
Sanitary Ware	178
Electrical Ware	190
Industrial Ceramics	195

PART IV   LABOUR 201

10 WORKERS, WAGES AND PRODUCTIVITY 203

Employment	203
The Occupations of Men and Women	205
Skill	205
Low Pay?	209
Job Satisfaction and Working Conditions	211
Labour Turnover and School Leavers	215
Industrial Relations	220
Productivity	227

PART V   CONCLUSIONS AND FUTURE PROSPECTS 233

11 CONCLUSIONS AND FUTURE PROSPECTS 235

Employment	236
Substitute Materials	237
Demand Stability	238
Industrial Relations	240
Overseas Branches of British Firms	241

Competition and Monopoly   242
Past, Present, and Future   245

PART VI   APPENDICES   247

Appendix 1   Bibliography   249
Appendix 2   Statistical Sources and Tables   254
Appendix 3   Amalgamations and Takeovers in the Pottery   259
             Industry
Appendix 4   The Pricing of Tableware   266
Appendix 5   An Analysis of the Demand for Tableware   273

INDEX   287

Part I

COMPETITION, TECHNOLOGY AND
ORGANISATION

1

THE STRUCTURE OF THE POTTERY INDUSTRY

In 1970, 53 600 persons were employed in the pottery industry. Of these, 15% were engaged in making tiles, 11% made sanitary ware, 7% made electrical ware and 65% made domestic ware (tableware and ornamental ware); the remaining 2% made industrial ceramics, etc. Tiles, sanitary ware, electrical insulators and tableware have one thing in common—clay. Clay is a cheap raw material; it is also awkward to handle, so that experienced managers and workers are needed as well as ingenious processes and machines to achieve products that are both satisfying and satisfactory. It is usual for each group of products to be manufactured in separate factories and the markets served are quite distinct. Tiles and sanitary ware are building components, electrical insulators are components of electricity generating and distribution equipment and electrical appliances and machines. In contrast domestic ware is purchased for its own sake by households; and in addition cups, plates, etc. are purchased in substantial quantities by catering establishments. Industrial ceramics are widely used in the engineering and chemical industries.

The pottery industry is therefore highly diversified. It is organised to supply an extremely wide range of markets at home and abroad. Each sector of the industry tends to have specialised managers and workers. There are even separate Trade Associations. The mobility of labour tends to be much greater within each sector than between sectors. Doulton & Co., which operated sanitary ware, electrical porcelain, industrial ceramics and domestic ware factories even sent a technical manager on a fact-finding mission to discover if improvements made in making one group of their products could possibly be utilised in others.

The diversity severely limits the number of general statements that can be made about the industry as a whole. Each sector of the industry must be considered as if it were an industry in its own right. For example, it does not make sense to discuss mechanisation in the pottery industry as if the industry were homogeneous. It makes sense, however, to discuss the high degree of mechanisation achieved by the manufacturers of tiles and its virtual absence in sanitary ware and large insulators. Also the contrast between the high degree of mechanisation achieved by manufacturers of relatively cheap earthenware and the limited amount of mechanisation present in some fine china factories. Some idea of the different degrees of mechanisation achieved in the various sectors of the industry may be obtained from the figures contained in *Table 1.1*. In 1963 domestic ware accounted for 61% of employment and only 47% of output; in contrast tiles accounted for 23% of output and only 15% of employment.

Table 1.1   THE MAIN SECTORS OF THE POTTERY INDUSTRY

Sector	1963* Output (£'000)	(%)	1963* Employment (No.)	(%)	1970 Employment (No.)	(%)
Domestic ware	35 781	46·8	35 648	61·4	34 800	64·9
Tiles	17 164	22·5	8 651	14·9	8 100	15·1
Sanitary ware	10 588	13·8	5 418	9·3	5 600	10·4
Electrical ware	10 239	13·4	7 061	12·2	4 000	7·5
Other	2 563	3·5	1 269	2·2	1 100	2·1
Totals	76 336	100·0	58 047	100·0	53 600	100·0

Sources: *Census of Production*, Board of Trade: H.M.S.O., London (1963); and *Pay and Other Terms and Conditions of Employment of Workers in the Pottery Industry*, National Board for Prices and Incomes, Report No. 149, H.M.S.O., Cmnd. 4411, London (July 1970).
* The 1963 figures cannot be compared directly with the 1970 figures for in 1963 firms employing less than 25 persons were excluded.

OUTPUT AND EMPLOYMENT

Figure 1.1 shows 50 years of output and employment in the pottery industry. The output trends up to 1948 involve some guesswork, mainly because firm data for price movements are not available for the earlier period (details of the calculations are contained in Appendix 2). *Figure 1.1* also shows that in the 1920s the pottery industry contracted and, in addition, it suffered from severe fluctuations in output and employment. Over this period the rate of profit on turnover was only 4% and turnover and capital were approximately equal.* Along with other staple British industries such as wool, cotton, coal and shipbuilding, the pottery industry suffered from a loss of export markets and a failure of the home market to take-up the slack; the rate of profit reflects how the industry suffered from excess capacity.

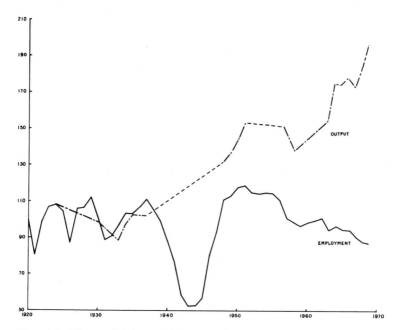

Figure 1.1   The pottery industry 1920–1969: Index numbers of output and employment
(1930 = 100)

A feature of *Figure 1.1* is the severe fall in employment between 1940 and 1945; output, of course, contracted too in war-time (output figures for the war-time years are, however, not available and the regular broken line on the graph is simply for purposes of continuity).

* Board of Trade, *Pottery* (Working Party Report), H.M.S.O., London (1946).

In the interests of the war effort the industry was ruthlessly curtailed. This led to a substantial boom from 1945 to 1955 when the loss of war-time production for both exports and the home market was made good. The war and the post-war shortages meant that a necessary and radical transformation of the industry was postponed; after the 1955 recession, the reconstruction of the industry, which involved the elimination of many small firms, was a testing time.

In the immediate post-war years many small firms entered the industry, made handsome temporary profits and cleared-off after 1955. Reconstruction was difficult because the 1920s and the 1930s had been lean years and firms tended to depend on retained profits to finance investment. Buildings and equipment were antiquated and unsuited to the post-war conditions of scarce labour and high wages. In addition, the owners and managers were forced to learn new methods and adopt new attitudes rapidly. This was not easy in an industry which had changed only gradually over a period of 200 years. One important aspect of the transformation was the substitution of large firms for small ones. The trend is clearly apparent in *Table 1.2*.

Table 1.2 NUMBER OF FIRMS, 1935–1970*

Sector	1935†	1958	1963	1970‡
Domestic ware	230	116	94	85
Tiles	50	25	18	14
Sanitary ware	30	14	14	12
Electrical ware	15	12	13	12
Totals	325	167	139	123

Source: *Report on the Census of Production*, Section 103, (Pottery), H.M.S.O., London (1963). *Firms employing 25 or more persons. They accounted for 96·4% of gross output in 1963. †The 1935 figures are estimates based on the *Census of Production* data for establishments. ‡The 1970 figures are based on take-overs, liquidation, etc. since 1963.

Output increased substantially between 1935 and 1970 yet the number of firms fell from approximately 325 to approximately 123. The fall in numbers is partly accounted for by amalgamations which have been, and continue to be, a feature of the industry. The fall in the number of firms has been accompanied by a fall in employment. *Figure 1.2* shows a steady fall in employment and an increase in real output since 1948; obviously labour productivity has increased substantially since 1950. Labour productivity increased in three main ways:

7

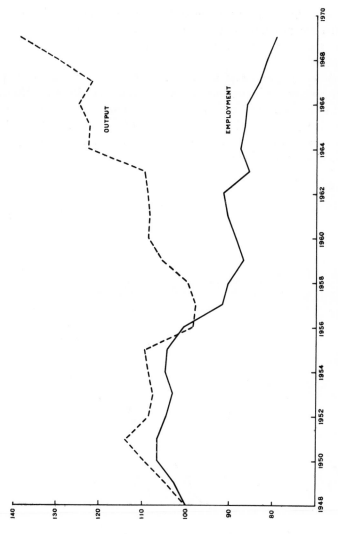

Figure 1.2   The pottery industry 1948–1969: Index numbers of output and employment (1948 = 100)

1. By the mechanisation of many processes, including the transportation of materials and ware between processes.
2. By improvements in body preparation and firing methods.
3. By the rationalisation of product ranges.

These three changes resulted from a substantial and sustained investment programme, the concentration of production into larger units and, most important, a shift from intuitive to scientific management.

Figure 1.3 shows that in the early 1960s the increases in output were concentrated mainly in tiles, sanitary ware and electrical porcelain. However, after 1967 the increases in domestic pottery output (devaluation provided a big stimulus to exports) tended to offset a lack of growth in electrical porcelain and sanitary ware caused by a general stagnation of the economy.

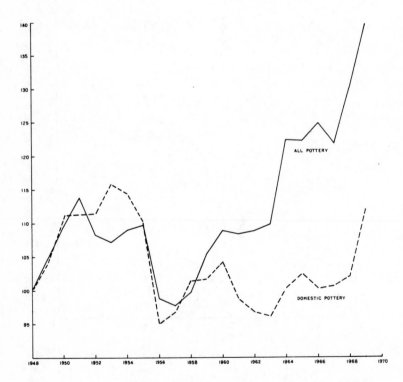

Figure 1.3   The pottery industry 1948–1969: Index numbers of production
(1948 = 100)

The pottery industry is a small industry. It accounts for less than 1% of the output of all manufacturing industries. We have estimated that between 1907 and 1963 the output of the pottery industry increased by some 30% whereas the output of all manufacturing industries more than trebled over the same period. Whereas the pottery industry had reached maturity by the end of the nineteenth century, manufacturing industry as a whole in Britain increased substantially because of the rise of new industries, such as vehicles, chemicals, electrical engineering, electronics, etc. which more than offset the decline in coal, cotton, etc. *Figure 1.4* contrasts the modest growth of

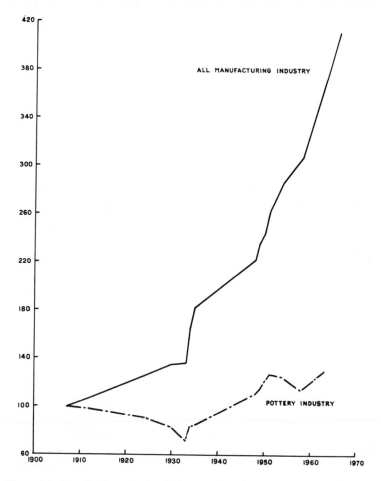

Figure 1.4   Comparative growth of the pottery industry and all manufacturing industry 1907–1968

output in the pottery industry since 1907 with the much higher rate of growth achieved by all British manufacturing industry. The growth achieved by the pottery industry in the 1960s is more· impressive when viewed against the long period of stagnation which had dominated the industry since the 1890s. We must, of course, be aware of projecting the growth of output in the 1960s too far into the future for it is possible that the industry has reached a new plateau and will need to struggle hard to avoid contraction.

Table 1.3 contains data from the 1970 Census of Production and comparable data for 1963 and 1968 which were also Census years. The figures provide the definitive record of changes of the industry as a whole in the 1960s.

Table 1.3 'POTTERY INDUSTRY: GROSS OUTPUT, NET OUTPUT AND EMPLOYMENT

Year	Gross output (£m.)	Net output (£m.)	Net output per head (£)	Total employment* ('000)
1963	79·7	51·3	848	60·4
1968	104·2	66·7	1 200	55·6
1970	118·6	73·2	1 462	50·1

Source: Trade and Industry (23 December 1971) (preliminary Census Returns for 1970).
*Includes working proprietors.

EXPORTS

The British pottery industry is highly dependent on export markets. Manufacturers of good quality domestic ware tend to regard the home market as just one market among many. Some 60 years ago approximately one third of the total output of the industry was exported. Today the proportion is only slightly less. Approximately 50% of the output of the domestic sector of the industry is exported. Exports account for approximately 25% of the output of tiles and for some 20% of both electrical porcelain and sanitary ware. Relative to its size the pottery industry makes an impressive contribution to the balance of payments. Its grip on the home market is such that imports are not large in spite of relatively modest tariffs. Also the import content of the pottery made in Britain is negligible. Table 1.4 provides a summary of the industry's exports in 1970 and 1971.

THE STRUCTURE OF THE POTTERY INDUSTRY 11

Table 1.4 EXPORTS OF POTTERY FROM THE UNITED KINGDOM (£M.)

Exports	1970	1971
Porcelain and bone china	10·4	12·9
Other types of household ware*	16·6	19·4
Ornamental ware	1·7	—
Glazed tiles	4·3	4·3
Electrical ceramics	1·9	2·5
Sanitary ware	4·0	4·4
Industrial ceramics	0·8	0·6
Totals	39·7	44·1

Source: *Overseas Trade Statistics of the United Kingdom* and *Business Monitor*, (Pottery) p. 6. (1971 and 1972).
*Mainly earthenware.

FACTORY SIZE

The size of the average pottery factory has been increasing. If we exclude establishments employing less than 25 workers, the employment per factory in 1935 was 190. It rose to 249 in 1958 and to 266 in 1963.* This means that the size of factory, measured by output, increased to a much greater extent because capital per worker also increased over the period and technical improvements were also introduced. In 1963 the largest factories were in electrical ware (average employment 321) and sanitary ware (average employment 319). The average employment per factory in tiles was lower (288) and lowest of all was average employment per factory in domestic ware (268).† Over the period 1935–1963 technological changes favoured larger production units; also larger firms tended to abandon their smallest factories and concentrate production in their larger units.

Large firms in the pottery industry tend to operate a number of factories rather than attempt to concentrate output into one large factory. It is unlikely that a smaller number of much larger factories would necessarily be an improvement in the context of the pottery industry. Medium-sized factories suit localised labour markets in North Staffordshire and they contribute towards easing travelling-to-work problems. They also, possibly, contribute towards better in-

* *Census of Production*, H.M.S.O., London (1963).

† The average employment quoted in all factories, 266, is below the average of domestic ware factories, 268. This arises because the average for all factories includes 'other establishments'.

dustrial relations. New factory buildings are extremely expensive, so firms tend to prefer to renovate old premises. There are also no fixed items of plant which would give rise to economies of scale, provided a minimum size is achieved, if output were concentrated in larger factories. (Large factories tend to operate a number of tunnel ovens.) It is a basic proposition of this book that there are economies of scale which operate when firms increase in size. However, they are based on marketing, commercial and financial economies, as distinct from production economies. Hence, in the pottery industry, large firms find it economic to operate a number of factories.

EXTERNAL ECONOMIES

We have noted that the labour force of the pottery industry was virtually the same size in 1920 as in 1960 and that employment contracted after 1951. Part of the fall in employment, however, may be explained by processes that at one time were performed within the industry being performed elsewhere. In the twentieth century the pottery industry has become more specialised. Fewer pottery firms now grind their own flints, bones or colours or make their own glazes or lithographs. In addition, they expect specialists to build their kilns and design and install their machines. Over the years the pottery industry has become much more dependent on non-pottery firms in supporting industries. These developments have greatly strengthened the industry. The supporting industries cater for many other industries as well as the pottery industry. The technology of the kiln makers or the machine makers has been strengthened by their need to supply many industries with kilns and machines and not just the pottery industry. Similarly with colours or lithographs or packing materials, the industry has benefited from improvements in technology which were geared to a much wider market than the pottery industry alone could provide. The activities of the Devon and Cornwall clay producers provide an example of how the pottery industry has benefited from developments in technology and organisation external to the industry.

English China Clays Ltd. is an international company. It operates as a holding company and its subsidiaries mine and treat china and ball clay, as well as operating in the building industry, in quarrying, transport and engineering. In china clay it accounts for some 20% of world production (communist countries excluded). Approximately 20% of china clay production is consumed by the pottery industry.

In contrast, the paper industry absorbs some 70% of the output. In 1970 the turnover of the clay division was £41 million, of which exports (£29 million) and overseas production amounted to £31 million. An important aspect of this giant company's development has been the treatment and standardisation of its china and ball clays. Pottery firms can now place orders for clays which are delivered ready for immediate use and with future deliveries guaranteed of a standard quality and consistency. The second largest producer of china clay and the largest producer of ball clay in Western Europe is Watts, Blake, Bearne & Co. Ltd. The sanitary ware sector of the industry has benefited in recent years by the development of calcined clays by this company which help to speed up firing schedules. These large and specialised companies have enabled pottery manufacturers to reduce substantially their raw material preparation activities. Also they have possibly kept down the cost of their requirements.

The relatively small size of the British pottery industry renders it uneconomic for suppliers to specialise only in its requirements. They therefore produce machines and equipment which, with modifications, can be used in many industries. Thus the pottery industry can expect to have to contribute very little towards the development costs of its requirements. Not only are prices to the pottery industry kept down, but the industry also benefits substantially from the improvements in technology which are pioneered by its suppliers. A long-established machine manufacturer to the pottery industry is William Boulton (Holdings) Ltd. of Stoke-on-Trent. After 1960 this firm decided, for the sake of profit, growth and security, to restrict its output for the pottery industry to relatively heavy machines for materials preparation. Other firms supply making and decorating machines, etc.

All the machines now manufactured by William Boulton can now be modified for use in other industries. For example, Boulton filter-presses treat, in addition to clay, sugar, starch, yeast, sewage sludge, chemicals, works effluent and paints. Their mixing and blending equipment can be used to process insecticides, drugs, flour and powdered metals, in addition to flints and clay. William Boulton has also taken over 10 other engineering firms to extend the activities of the Group. The recent drastic changes in the firm must have resulted in pottery manufacturers being able to purchase better and cheaper machines.

THE POTTERIES

In 1938 half the working force of Stoke-on-Trent was employed in the pottery industry and the employment opportunities of men and women and boys and girls in Stoke-on-Trent were dominated by the pottery manufacturers. Since 1938 the industrial structure of the area has become more diversified with the growth of engineering, other industries and services. Now approximately one third of the labour force is employed by the pottery industry. In North Staffordshire, which comprises, Stoke-on-Trent, Newcastle-under-Lyme and neighbouring villages and rural areas, one-fifth of the labour force is in the pottery industry. In 1963 approximately 82% of the labour force engaged in the manufacture of pottery was located in North Staffordshire (see *Table 1.5*).

Table 1.5 THE LOCATION OF THE POTTERY INDUSTRY, NORTH STAFFORDSHIRE AND ELSEWHERE

Sector	N. Staffs.	Employees in 1963 Great Britain	N. Staffs. as % of G.B.	1956
	('000)	('000)	(%)	(%)
Domestic ware	29·5	31·5	94	93
Tiles	4·9	7·2	68	77
Sanitary ware	1·9	4·3	44	42
Electrical ware	3·7	6·0	62	57
Totals	40·0	49·0	82	77

Sources: *Census of Production* (1963), *Pattern for Progress*, H.M.S.O.; *the Second Report of the Joint Standing Committee for the Pottery Industry*, Appendix I, (1970); and *Industrial Health: A Survey of the Pottery Industry in Stoke-on-Trent*, H.M.S.O., London (1959).

*The percentages are merely approximations. The figures for North Staffordshire and Great Britain were not collected on the same basis.

Why is the industry so concentrated in Stoke-on-Trent? The original attraction was the coal produced from the North Staffordshire coalfield, which was particularly suited to firing ware. It was much cheaper to transport clays to Staffordshire from Devon and Cornwall than to transport coal from Staffordshire to the clay deposits. Now that coal is no longer used for firing and fims use instead gas, electricity or oil, presumably firms could be located in any of Britain's numerous manufacturing areas. However, there are good reasons why pottery firms stay in North Staffordshire. It would be exceedingly difficult for a large firm to recruit a labour force away from Stoke-on-Trent.

Small firms have gradually grown well away from the Midlands in recent years; nevertheless, some of them find it difficult to sustain their labour force. The North Staffordshire Polytechnic specialises in training managers for the industry and the area contains a bewildering variety of firms which service the industry—kiln builders, machine makers, lithographers, colour suppliers, raw materials suppliers, transport, finance and insurance companies, etc. The British Ceramic Research Association, which undertakes basic research into ceramic materials and processes, is located in Stoke-on-Trent. The headquarters of the various employers' and trade associations, the Potters' Club and the Ceramic and Allied Trades Union are also in Stoke-on-Trent. Some well-known and successful pottery manufacturers operate away from North Staffordshire. They include the Royal Worcester Porcelain Company, Crown Derby, Joseph Bourne (Denby), and Poole Pottery. Other firms too, could succeed away from North Staffordshire if they wanted to. However, the pride in and loyalty to their industry of the men and women who work on the Potbanks is a unique asset of North Staffordshire.* An important reason why the industry continues to be concentrated in North Staffordshire is that there are many more profitable manufacturing activities to be undertaken away from the area than pottery manufacture. Inertia too has played its part; once firms find congenial situations, why should they change?

Table 1.5 indicates that sanitary ware, tiles and electrical porcelain are concentrated in North Staffordshire to a lesser extent than the manufacture of domestic ware. The manufacture of tiles dispenses with most traditional pottery skills and this encourages dispersion. If sanitary ware production were not dispersed, delivery charges would increase considerably: the ratio of value to weight of a hand-basin is considerably less than that of a dinner plate. Many of the manufacturing processes in electrical ware and industrial ceramics are essentially engineering techniques applied to clay rather than metal. So the labour force of North Staffordshire is less of an attraction than it is for the manufacturers of tableware.

COMPETITION

The pottery industry is, and for a long time has been, dominated by competition. Competitive forces have been responsible for the

---

* Mervyn Jones in *Potbank* (Secker & Warburg, London 1961), emphasises the uniqueness and pecularities of Stoke-on-Trent and its citizens. The author neglects, however, to emphasise the development of modern industry and new and less idiosyncratic attitudes. (The 'Potbanks' are the pottery factories and the term originates from the fact that many of the early factories were located on banks of clay.)

structure of the industry and changes in the structure. The argument of Chapter 3 is that competition results in the survival of a few large firms which increasingly dominate the industry. We do not consider it useful or even desirable to spend much time deploring the loss to the industry and the nation of the small family firm. They had a good innings. However, a number of small firms remain in business, particularly in the domestic sector of the industry, and reasons will be given why they are likely to survive. Part II of the book is concerned exclusively with the domestic sector of the industry; it includes chapters on production costs, home demand and exports. Part III is concerned with sanitary ware, tiles, electrical insulators and industrial ceramics. Part IV is concerned with pottery workers and their working conditions and earnings, etc. Labour problems are approached from the point of view of the industry as a whole. Finally, in Part V we indulge in speculation. It is mainly concerned with the effects on the size and structure of the industry of changes in technology and changes in demand. In 30 years time will china, earthenware and stoneware still be used by every household and most catering establishments for tableware? Will they be replaced (or merely supplemented) by plastics, toughened glass, stainless steel or, Heaven forbid, impregnated paper? Will plastics replace porcelain in public and private conveniences? Will wallpaper, paint and/or fibre boards replace ceramic tiles on walls? Will developments in pottery firms abroad, including overseas subsidiaries of British companies, undermine British overseas markets for pottery? Of course the changes in future demand will be affected substantially by technological changes. An important factor which has sustained the sales of ceramic tableware in the 1950s and 1960s has been the cheapness of pottery relative to other materials; this situation may be reversed by technological changes. Fortunately, from the point of view of pottery manufacturers this is a somewhat unlikely possibility. What is a strong possibility however, is improvements in technology permitting output to be sustained with a severe cut in the size of the labour force. This is something which occurred in the 1960s in the tiles sector of the industry. In the 1970s the other sectors of the industry are likely to follow the lead of tiles and as output increases the labour requirements of the industry are likely to contract considerably.

SUMMARY

In this first chapter we have noted that the pottery industry consists of five fairly self-contained sections which produce widely differing products and which serve quite different markets. The age of the industry was noted and the extent to which it suffered from stagnation between the two World Wars. In recent years, however, we noted how the industry has been rejuvenated. Its firms and factories have increased in size, firms have become more specialised and output and labour productivity have increased. Two characteristics of the industry have persisted throughout the bad and the good years—its high dependence on export markets and its concentration in North Staffordshire. In Chapter 2 we explain how pottery is manufactured and note how technology and production methods have changed in recent years.

2

TECHNOLOGY

The first potters used clay very much as they found it. Since natural clays are of highly variable composition, such raw materials could not be relied on to behave in anything like a uniform way during shaping, drying and firing; nor could they be guaranteed to give finished wares of predictable quality. One important line of advance, obvious in retrospect though probably not so in the beginning, was to select, blend and purify clays, and mix other ingredients with them, to produce bodies with better controlled and more nearly constant properties. By Roman times progress along these lines had reached such a pitch that Samian ware, produced in widely scattered parts of the Empire, showed little or no appreciable difference in the composition of its reddish-brown body from one place to another. Strict control of body composition is still one of the ceramist's prime concerns, and he calls on any resources of science and technology that can help him to achieve it.
Maurice Chandler, *Ceramics in the Modern World*, Aldus Books Ltd., London (1967).

The pottery industry is not part of, say, the chemical industry or the engineering industry, because clay imposes severe technical limitations on manufacturing processes. Ceramics are different. It is important that a grasp of the elements of production processes is obtained, otherwise the reader may regard certain hand processes or even semi-automatic processes as backward or old-fashioned, whereas they may be entirely appropriate, at ruling wages, for the 1970s. Of course, a lot of problems would be solved if plastics or other less complicated materials were substituted for clay. However, clay is plentiful and cheap and the economics of the situation render it extremely difficult for substitute materials to take over from clay.

MATERIALS

Making pottery may be compared with making biscuits. Biscuits come in a great variety of shapes, textures, colours and sizes, plain and fancy, and they have one common ingredient—flour. Similarly, with pottery the variety is enormous, the one common ingredient here being clay. In addition, both biscuits and pottery are manufactured by the application of heat. Because of the much higher temperatures involved, pottery is fired rather than baked. Some biscuits may dispense with flour in their manufacture and yet they are conveniently defined as biscuits. Similarly, a few pottery products do not contain clay, instead alumina, silicon nitride or tungsten carbide may be used to produce 'special ceramics' or 'engineering ceramics'. An enormous range of pottery products is achieved by varying the ingredients, types of clay and various materials mixed with the clays, in addition to varying the making, glazing, decorating and firing processes. It is the aim of the professional potter, in contrast to the art potter, to standardise every ingredient and process to achieve consistently a standard and unvarying product. In recent years the substitution of scientific attitudes for craft attitudes has rendered this objective much easier to achieve.

CLAY

Clay deposits are widespread and plentiful. It is a material which can be shaped easily, it remains rigid when dried and it acquires considerable strength after it has been fired. After it has been formed, and before it has been fired, clay regains its elasticity when it has been thoroughly wetted. This means that only before firing commences can faulty ware be reconstituted back into clay. There is a satisfaction to be gained by working with clay or using clay products which derives partly from the age of the material. Some clay deposits were laid down 1000 million years ago. In addition, clay has been used as an industrial material for thousands of years. Ball clays are included in pottery recipes to provide plasticity to the body; they provide a high 'green' strength which enables ware to be carried about prior to manufacture. In contrast, china clays which add whiteness to the bodies are not nearly as plastic as ball clays. Whiteness is a quality of domestic ware products which has been eagerly sought after by British pottery manufacturers since the eighteenth century.

RECIPES

The traditional recipe for an earthenware body is:

Ball clay	25 parts by weight
China clay	25 ,, ,, ,,
Flint	35 ,, ,, ,,
Stone	15 ,, ,, ,,

Flints are calcined, crushed and ground to a thick suspension. Flint is a non-plastic material which adds whiteness to the body and renders it easy to dry. The quantity used and the method of its use largely determine the crazing* resistance of the ware. It is used as a controlling factor for matching the expansion of the body and the glaze. The Cornish stone of the traditional earthenware body is a granite with a high feldspathic content. The stone has to be crushed and ground fine before it is added to a mixing in the form of a suspension that has the consistency of a thick cream. The amount of stone used determines the degree of vitrification and the range of temperatures over which the vitrification of the body takes place.

Josiah Spode (1754–1827) developed bone china. The recipe was approximately as follows:

Bone ash	52 parts by weight
Cornish stone	24 ,, ,, ,,
China clay	24 ,, ,, ,,

Because bone ash, unlike calcined flint, provides plasticity, ball clay may be dispensed with in the manufacture of bone china. A little

* Crazing occurs if the thermal expansion properties of a glaze and the body to which it is attached are not properly matched. Since glazes are weak when subjected to tension and relatively strong under compression, the thermal expansion of the glaze should be less than that of the body, so that when cooling takes place after glost firing, the tendency is for the glaze to contract less than the body. The contraction of the two must in fact be the same since they are attached to each other and the overall result is therefore that the glaze is under compression. If the reverse is true, the glaze is under tension, and likely to 'craze', i.e. show a network of cracks.

ball clay, however, is usually added to increase plasticity. A bone china body is the strongest of all domestic ware bodies (used sparingly it can suggest extreme fragility); it has a high degree of whiteness and, for good measure, it is translucent.

MANUFACTURE

Earthenware bodies are normally first-fired at a temperature of about 1150°C and they have a porosity (the proportion of air spaces in a material to air spaces plus solid particles) of 10–15%. Bone china bodies are first fired at a temperature of about 1250°C. At this temperature, wares are on the threshold of being molten and to retain their shape are frequently supported in the kiln. The porosity of bone china is very low.

Hard porcelain, which is mainly manufactured in Britain by Worcester Royal Porcelain Co. Ltd., consists of about 50% by weight of china clay and, in addition, quartz and feldspar and a small quantity of ball clay. Its initial firing temperature is relatively low: 1000°C. Vitrification occurs subsequently in a glost kiln, after it has been glazed, at a temperature of 1400°C. It, too, has a low porosity and it is translucent. The hard porcelain bodies which are used in the manufacture of sanitary ware and electrical ware are similar to those used in the manufacture of domestic ware.

Stoneware contains only small quantities of china clay and a very high proportion of ball clays. It is first-fired at a temperature somewhere between 1150°C and 1250°C. It is considerably less porous than earthenware, being a much more robust material. It tends to be grey or brown rather than white, and it is not translucent.

STAGES OF PRODUCTION

Figure 2.1 displays the series of processes which constitute the manufacture of domestic ware. It may be seen from the figure that ware may be formed by a dust pressing process, as is usual in the manufacture of tiles; it may be formed on a wheel or it may be cast. The towing and fettling processes* rub off rough edges after ware has been

---

* Towing and fettling are the processes used to remove the rough edges from pieces of ware prior to biscuit firing. The surplus material is removed using an abrasive wheel or belt, and before the introduction of strict dust-control regulations, workers engaged in these processes ran a high risk of contracting respiratory complaints.

dried, and the first firing results in biscuit ware. Ware may be decorated under-glaze or on-glaze. The firing which vitrifies the glaze is known as the glost fire, and the firing which fixes on-glaze decoration is the enamel fire. Elaborate decoration may be applied in a number of

Figure 2.1   The manufacture of domestic ware

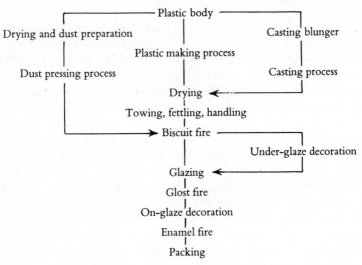

Figure 2.1   The manufacture of domestic ware

separate stages, each one being fired before the next is applied. Finally, ware is given a final inspection and is then packed for despatch. It is possible to glaze dried ware before it is fired. This permits once-fired ware to be manufactured. This speeds up the production process considerably and contributes towards low costs and prices. Firms which manufacture it would claim that it is not inferior to twice-fired wares. It is normal, however, for ware to be subjected at least to a biscuit fire and a glost fire. Sanitary ware is normally cast and large pieces of electrical ware are formed by turning on a lathe. In recent years an important innovation was the discovery of how to make sanitary ware needing only one firing. The outcome is perfect pieces of hard porcelain. In industrial ceramics, since the 1950s, new bodies and making techniques have been developed to provide components manufactured to more accurate dimensions and greater reliability. We will now describe each stage—body preparation, making, drying, first-firing, glazing and decorating and additional firings—in more detail.

BODY PREPARATION

Figure 2.2 shows how an earthenware body is prepared. It is now customary for materials to be supplied to the manufacturer according to standard specifications and in regular particle sizes. This is in contrast to what happened only some 20 years ago, and still happens in some instances, when great lumps of clay and stone had to be broken down and treated before being mixed together. Now specialised millers supply most manufacturers with their requirements and obtain and

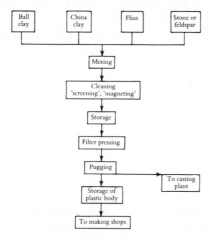

Figure 2.2   The preparation of an earthenware body

pass on the benefits of economies of scale in the process. Blunging is a term used when solids are being dispersed into liquids to form a suspension or slip.* New types of blungers have been developed and introduced which considerably reduce the time required for the production of consistent slips. The slip permits unwanted materials to be removed relatively easily. Magnets are used to eliminate iron particles which, by their presence in a slip, would discolour and blemish finished products. The next step is to remove surplus water by filtration under pressure. Not so long ago filter pressing required

---

* Slip is a suspension of clay particles in water. Mixing slips of different types has been found to be the best way of obtaining a body of uniform composition (the alternatives are mixing different clays in the plastic state and mixing the dry ingredients). Slip can be converted to plastic clay by removing most of the water, which is done using a filter press. Cast articles are produced by pouring slip into a plaster mould, which absorbs water from the slip and causes a film of clay to be built up on the inside of the mould.

long hours of hard manual labour. Now presses which drop the slabs of filtered clay onto trucks or moving belts have rendered working in a slip house less severe on human muscle. Before the body can be used all air must be removed and this is achieved by means of a pugging machine. It is a large mincing machine which shreds the clay, takes out the air in a vacuum chamber and then reconsolidates the clay into a solid plastic slug which is extruded from the nozzle of the pug. Pugging machines too have been considerably improved in recent years. The composition and consistency of the clays are controlled to suit various types of making machines, glazes, colours and firing times. One difficult problem with mechanising the making processes has been that each firm has tended to prefer its own peculiar body; the result has been that it has been impossible for the machine makers to provide the industry with mass-produced and standardised machines. This situation could be remedied if a greater degree of body standardisation were agreed upon by manufacturers. Takeovers and amalgamations have contributed towards this objective. A desirable development would be for a few standard bodies to be made centrally and delivered to the various factories.

A visitor to a pottery is first of all conducted around the slip house. There, in an atmosphere of cool dampness he sees hoppers full of clays and stones and flints and tanks filled with slip and pumps and pipes, as well as the blungers and pugging machines. He is inclined to be impatient, after ten minutes of contemplation, to see the making and decorating shops and experience the warmth of the kilns. His guides, however, know about fundamentals and keep him there longer; they have learned the hard way that the basis of good potting is carefully prepared and consistent bodies. So the slip house must be admired before the visitor is permitted to advance to see the clay being formed, hardened and glazed.

Slip for casting operations is prepared by mixing clay from the filter presses or from the pug mill with water and deflocculants. Deflocculation enables a thick slip to flow as readily as a slip which contains much more water. A slip which contained too much water would never build up an adequate wall thickness and moulds could not then be used to make large pieces as in the production of sanitary ware. In tiles dry dust is used instead of plastic clay in the making process. Slip is pumped into a spray drier which produces free flowing dust from spray droplets that have been rapidly dried in a current of hot air. This method, which has recently been introduced, is a marked improvement on taking the clay from the filter presses, drying it and then grinding it to dust.

MAKING

The shapes which will be produced are the ultimate responsibility of top management. Their agent will be the designer, who may be a director, and he will work closely with a modeller, and both in turn work in conjunction with a mouldmaker. One of the problems facing the modeller is scaling-up as ware contracts significantly during its first firing. The ware is produced with the aid of moulds and profile tools attached to machines and designed for repetitive processes.

The art of forming plastic clay into three dimensional shapes is one of straining the clay in such a manner during making that the structural changes that take place do not cause the article to go out of shape during firing. In other words, clay must be worked-up properly to its final shape, if the clay should be abused or ill-treated in the forming process then it is liable to distort or crack in the kiln. Large and awkwardly shaped pieces such as coffee pots or sauce boats tend to be made by shaping in plaster of paris moulds; however, the bulk of the ware is now made either by jollying or jiggering. Both processes achieve symmetry by means of rotary motion on a wheel. Moulds and profile tools are substituted on the machines for the hands and fingers of the craftsman potter. Jollying refers to the making of hollow ware,

Plate 2.1   Casting a coffee pot (Paragon China: Royal Doulton Tableware Ltd.)

such as bowls and cups, and jiggering refers to flat ware such as plates and saucers. In jollying, the mould gives the ware its external shape and a profiling tool provides the internal shape. On a semi-automatic flat making machine a slug of pugged clay is pre-formed to a flat pancake on a batting out machine. The operative throws the pancakes accurately and firmly onto the surface of the mould, with which it is essential that firm contact be made to prevent air being trapped between the clay and the mould. After the plate has been jiggered it is trans-

Plate 2.2 Plate making (Royal Crown Derby Porcelain Co. Ltd.)

ferred to a drier still on its mould. Machines are now operating which feed themselves automatically with slugs of clay and finally feed the formed ware into a drier.

Large sanitary ware pieces are made in plaster-of-paris moulds and automatic casting plants have been devised. Large electric insulators can be built-up by attaching together, with clay slip, jiggered and jollyed pieces. In tiles, where the pieces are small and the shape extremely simple, dust is forced into shape in high-speed stamping machines. Die stamping is also used for many types of small industrial ceramics, here alumina or some other material may be substituted for clay and shapes more complicated than tiles are obtained. The making processes are completed by drying and fettling. Drying is both a sophisticated and a complicated process. Waste heat from the kilns may be used for the process. A vast range of driers for specialised wares are now available on the market for manufacturers to choose from. In electrical ware and sanitary ware factories, where the pieces

Plate 2.3 Cup casting (Royal Albert: Royal Doulton Tableware Ltd.)

are large, large areas are given over to drying and moisture is gradually removed either by heating or by the constant circulation of cool air. If drying is not properly carried out, then water trapped in the body could cause pieces to explode in the kiln or in other ways damage the ware. After drying, ware is fettled to remove surplus clay and smooth rough edges and surfaces. Machines have been devised which speed-up fettling considerably. The formed, dried and fettled 'green' ware is now ready for firing in a biscuit kiln.

BISCUIT FIRING

Biscuit kilns are large pieces of equipment which, to justify their high cost, must be kept in operation at a high level of utilisation, day and night and at weekends for years on end. They are long brick tunnels through which truck loads of ware slowly travel on rails, the rails and wheels being in a comparatively cool, sealed-off portion of the kiln. The temperature gradually builds up towards the centre of the kiln where the ware is kept for a few hours at its maximum temperature and then it cools off again at the exit where the trucks are unloaded. Great care is taken in loading the trucks to ensure that obstructions

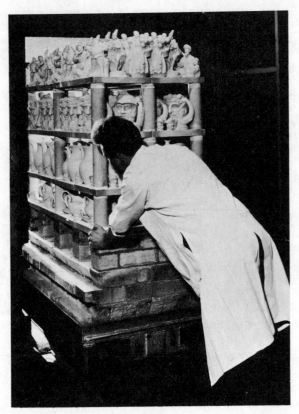

Plate 2.4   A loaded truck entering a gas-fired tunnel kiln (Doulton & Co. Ltd.)

are not created within the kiln. Also, if ware is not placed correctly it can become crooked and care must be taken to avoid large dust particles adhering to the ware. It is the capacity of the biscuit kilns which, in large measure, governs the short-run production plans of particular firms. Because of steady improvements of firing techniques, burners and fuels most firms have considerably increased the capacity of their existing tunnel kilns. It is not essential, however, to use only large tunnel kilns for biscuit firing. Small tunnel kilns may also be used or, instead, intermittent kilns* may be installed.

In the 1950s it would have appeared reasonable to predict that all

Plate 2.5   New 'portable' decorating kiln specially designed and built for Staffordshire Potteries Ltd. by Bricesco (the entire kiln and ancillary equipment stands above finished floor level)

new pottery kilns would be tunnel kilns. However, subsequent improvements in intermittent kilns and the desire to render firing

---

* An intermittent kiln is one which is not continuously fired. Firing of ware is thus carried out on a batch basis, the kiln being loaded with ware, heated to the required temperature, maintained there for the appropriate time and then allowed to cool. The traditional coal fired bottle-kilns were of course intermittent, and were largely replaced by continuously-fired tunnel kilns, fired by gas, oil or electricity, in the period 1935–1960. Since the war new types of intermittent kilns, relying mainly on gas and electricity as fuels, have been developed in an attempt to avoid some of the disadvantages (e.g. high capital cost and relative inflexibility in use) of tunnel kilns.

programmes less inflexible have resulted in intermittent kilns being preferred to tunnel kilns (this statement applies with greater force to glost kilns). It is likely that no new tunnel kilns will be installed in the British pottery industry in the future, except perhaps in tile and sanitary ware factories. Understandably, firms prefer to sustain a margin of their output over and above the capacity of their tunnel kilns by utilising intermittent kilns. A highly efficient 'top hat' kiln has been developed. Two refractory bases, A and B, are built on the factory floor close together. On A, whilst ware is being fired on B, ware is being loaded ready for firing. When firing on B is completed the cover of the kiln, complete with its heating elements, is raised by a crane and transferred to cover the ware on A. In this way firing can be virtually continuous if this should be required. Top hat kilns, however, can be switched on and off at will without undue expense and they can be operated on most fuels. This is a decided advantage as compared with tunnel kilns.

Firms have been experimenting in recent years with one-piece firing and some systems are already in operation. Truck loads of green ware are like large pieces of clay which must be heated to the centre. Hence firing times of 30 hours and over are common. In contrast, when pieces are fired one at a time the time taken can be only one minute or even less. With one-piece firing a kiln which included a drying-chamber could be linked to a making machine and making and firing would be combined in one operation. This would render the manufacture of pottery similar to processes in, say, textiles or light engineering. Hover kilns have been developed and are operating. The ware moves rapidly through the kiln on trays supported on a cushion of hot air. One problem has been coping with dust which may be deposited on ware by the disturbed atmosphere. The main consideration which retards the development of new firing techniques is the extremely low cost per unit which is obtained when large tunnel kilns are used at or near capacity.

It is usual for biscuit ware to be stored in a warehouse. This permits the making and glazing and decorating departments to operate without having to keep in step with each other. Firms which attempt to operate without a warehouse or an adequate one are apt to clog-up the passages and working spaces in the factory with piles of ware. Firms which produce once-fired wares are able to sustain a flow of production from the slip house to the despatch warehouse. By eliminating fetching, stacking and carrying operations production costs can be substantially reduced as compared with the more conventional twice-fired method.

GLAZING AND DECORATING

The porous biscuit ware is dipped or sprayed with a suspension of glaze which, when dried, covers the ware with a powder that will subsequently melt in the glost fire. Many glazing operations have been mechanised, the ware is either sprayed or it passes through a falling curtain of glaze which is kept circulating. In the glost kiln each piece must be kept separate, otherwise pieces will stick together when the glaze becomes molten. There is an enormous range of ceramic glazes and they may be thought of in terms of a layer of glass superimposed on the biscuit. Great care has to be taken to ensure that bodies and glazes are compatible. If there is any variation in the quality of the clays used this may require a change in the glaze or the temperature at which it is fired. This is only one example of the benefits which flow from raw materials being strictly standardised. If it can be arranged for a glaze to have a lower thermal expansion than the body of an appropriate amount it will be in a state of compression when cold and this can add to the strength of the ware. Insufficient compression is likely to result in crazing and cracking. Development work has substantially reduced the time taken for glost firing. Only a few years ago 30–40 hours was normal, now some 20 hours is usually sufficient time.

Before glaze is applied the biscuit may be decorated to obtain under-glaze decoration. Under-glaze decoration provides the advantage of extra protection for the colours. Because glazes require to be fired at higher temperatures than many colours, under-glaze decoration limits the range of colours that may be used. On-glaze decorations, in contrast, may be fired at lower temperatures which melt the fluxes in the colours while being insufficient to melt the glaze. Because different colours require different firing temperatures, pieces may be fired in enamel kilns—after they have been fired in glost kilns—a number of times. For the cheapest ware the amount of re-cycling through the enamel kiln is kept to an absolute minimum.

Decoration may be applied by hand. Hand decoration is important in the British pottery industry because customers at home and abroad are willing to pay extra for hand work applied to bone china, fine earthenware and stoneware dinnerware and ornamental wares. Much of the decoration on fine bone china, etc. is, however, applied by means of transfers. They are also used extensively to decorate the cheaper lines of bone china and earthenware. (Over the years the British pottery industry has had an unquenchable appetite for transfers of rosebuds, sprigs of lilac and hollyhocks. As intermediaries, the manufacturers have catered manfully for the undemanding tastes of the mass of the consuming public.) Colours and paper transfers are

Plate 2.6   Painting a figurine (Doulton & Co. Ltd.)

usually purchased from outside suppliers. The transfers are normally printed from lithographic plates using varnish rather than printing inks; the varnish picks up ceramic colours in dust form. The paper transfers are known as lithos and in 1963 the industry spent £1 million on them. The designs are transferred to the ware by pressing or rubbing after the lithos have been wetted. The designs are also transferred to paper from engraved plates. Recent developments permit colours and patterns to be stamped directly onto the ware by machines. Automatic offset machines are widely used. They transfer the design on an engraved copper plate by means of a parabola-shaped gelatin pad. This device permits designs to be transferred to irregularly shaped hollow ware. Developments in silk-screen printing have enabled ware to be mass produced with attractive traditional or contemporary designs. Finally gold and platinum are used to gild domestic and ornamental wares. Semi-automatic machines are now operating which band cups, saucers, bowls and plates. The metals are fired onto the ware in the same way as colours are fired in enamel kilns.

Plate 2.7 Automatic Murray Curvex operator (Ridgway Potteries: Royal Doulton Tableware Ltd.)

PACKING

Traditionally, after inspection and selection, finished ware was packed in straw or wood shavings in large hampers or barrels for despatch to wholesalers, retailers or importers. Ware continues to be packed in this way. However, firms are tending to favour impregnated paper board packs which are smaller and much easier to handle and which do not require messy packing materials. Now packs are not only lighter and more compact, they also can be made attractive with brand names and symbols which help to promote the sale of the contents.

SPECIALISATION

In this brief review of technology, four main areas of activity have been
identified: materials and body preparation, making, firing, and
decorating and glazing. It has been noted that the preparation of
materials, including colours, glazes and lithos, are mainly undertaken
by specialists on behalf of the pottery manufacturers. Because the
suppliers operate on a large scale and supply many industries, the
pottery manufacturers benefit from specialisation and not doing a
number of things for themselves which others can do better and at
lower cost. At present most British manufacturers operate slip houses,
making departments and decorating and glazing shops and operate
their own kilns. It would be possible for manufacturers to purchase
a few standardised bodies from specialist body makers. This is a
possible development in the industry in the years ahead. Also firms
could specialise in making biscuit ware and other manufacturers
could limit their activities to decorating and glazing. This would
not necessarily be a desirable development as internal administration
may be less costly than operating by means of buying and selling on
markets. A few firms have operated and do operate as decorators
and their high degree of specialisation permits them to be both small
and efficient. The developments which have been noted in firing,
glazing, making and decorating strongly suggest that small firms
and small factories, so long as there is specialisation, can be viable
in the domestic sector of the pottery industry.

Developments in technology, contrary to what many people appear
to believe, favour medium-sized rather than large-size units of production. It is now possible to select the machines and equipment and
plant which are appropriate to the chosen scale of operation. Entry
always has been relatively easy to the pottery industry and the recent
improvements in technology have not rendered entry more difficult.
These observations, however, would not apply to electrical ware,
sanitary ware or tiles. Although technological changes tend to favour
smaller production units and firms, the fact that a few large firms are
firmly established in the domestic sector of the industry must dis-
courage potential new entrants.

Most of the improvements in production methods have taken place
outside the pottery industry and have subsequently been made
available to it. Fuel provides a good example of this; the West Midlands Gas Board and the Midlands Electricity Board have been active
in developing kilns and burners and so have the oil companies.
Innovations made by suppliers of materials, kilns and machines have

been noted. One could go further and mention important but less tangible developments in office systems and equipment, marketing and financial techniques and wage payment and labour training schemes which have been adapted by the industry from the economy at large. As the economy at large has developed more efficient equipment and techniques so has the pottery industry changed and been modernised to conform more with its environment. Its own research personnel and work undertaken by the British Ceramic Research Association have contributed towards this end. It is to the credit of the industry that it has responded effectively to rapidly changing technology and changing attitudes. That the industry in most instances has reacted to stimulae from outside rather than initiated changes from within is not really surprising, it has been mainly a matter of relative sizes.

3

COMPETITION

The reasons for take-overs and mergers in this industry are many and the advantages to be gained equally numerous. These are not the monopolistic groupings of the giant cartels, but a genuine and necessary step forward in the industry's structure, permitting it to develop and use modern methods, to retain and improve its market at home and, of particular importance, its market and name overseas.
Arthur Bryan, Chairman, Wedgwood Ltd., Royal Society of Arts Lecture, November 1970.

The pottery industry developed because of the enterprise of family businesses, many of which were established in the eighteenth century. In contrast, the revival of the industry in recent years has been based firmly on the enterprise displayed by the managers of medium and large public companies.

FAMILY BUSINESSES

Family businesses were adequate basic units of the industry in the nineteenth century, mainly because capital requirements were modest. Skilled craftsmen would rent premises and gradually build-up a business with the help of brothers and sons, nephews and cousins. The chairman would probably arrange for finance and keep the books. One brother would act as works manager and the other would organise sales at home and abroad with the aid of salesmen and overseas agents. Families could supply adequate management so long as the industry was organised on a craft basis and technological expertise was restricted to a fairly superficial understanding of forming and firing clay. A cheap and plentiful supply of labour also contributed towards the success of family businesses.

Until 1945, North Staffordshire, as elsewhere, provided a more than adequate supply of cheap labour. On the basis of relatively low wages the employers demanded and obtained sustained effort and dedication to high standards of performance from their employees. The lack of an adequate number of alternative jobs in North Staffordshire meant that the employers could strike extremely hard bargains and hire and train labour in a haphazard way without retaliation from organised labour; in fact, the employees appear to have had a grudging respect for the toughness of the employers. The employers depended heavily on the skill and judgment of the workers to achieve standardised products in spite of variations in raw materials and in the temperature and humidity of the workrooms. In the factories the workers formed their own working groups. The craftsmen sub-contracted work from the employers and paid their helpers and assistants from their gross earnings. The employers operated mostly by rule-of-thumb methods; many of them could be regarded as merchants who sub-contracted manufacturing to their craftsmen. In sharp contrast today, the managers provide standardised materials in controlled conditions and the scope for judgment on the part of individual workers and teams of workers has been severely reduced. The de-skilling of labour has been accompanied by a substantial increase in the understanding and control of technical processes by managers.

Spode, Wedgwood, Minton and Doulton all introduced improvements which permitted the firms which bear their names to prosper over the years. Improvements were constantly made by family firms to body composition, glazes, moulds and drying and firing techniques. However, the improvements were not mechanical and few of them provided firms with marked advantages, in terms of lower costs, over the rest. Hence many firms survived and new firms replaced old ones. The basic unit of organisation in the industry after 1900 was the medium-sized firm employing from 300 to 1000 persons rather than smaller firms employing from 50 to 200 persons. Mechanisation had been introduced into the industry from fairly early on. In the 1860s mechanisation increased substantially. It took the form of transmitting power from one prime mover to a number of grinding machines, or to blungers and pugging machines or making machines.

All other operations, firing, decorating, glazing, packing, etc., were undertaken with relatively inexpensive tools and equipment in buildings which provided few comforts for workers or managers. Takeovers and amalgamations were frequent. Families would add factories to their existing ones as the size of the family increased, or sell-off factories as the size of the family decreased. The nature of take-overs changed considerably after 1955 and they quickly significantly changed the structure of the industry.

The following are some of the reasons why many family firms ceased to prosper after World War II and why they were replaced by fewer companies.

1. Tunnel ovens proved to be superior, in terms of cost, to bottle kilns. (In 1938 there were 2000 coal-fired intermittent kilns operating in the pottery industry, by 1970 they had all ceased to operate.) A tunnel oven often would require more land than was in the possession of a small firm on a cramped site. They were costly pieces of equipment which required to be utilised at near capacity on shift working to be profitable. The cost of their installation and operation were frequently beyond the capital resources of small businesses. Tunnels could not be installed leaving everything else equal. They involved firms in substantial supplementary expenditures on buildings and equipment.

2. Mergers tended to create a feeling of uncertainty and this triggered-off further mergers. Owners became convinced that it was essential to grow to survive in competition with larger groups.

3. The sharp increases in wages and in fuel and materials costs after 1946 meant that firms were forced to mechanise in an effort to keep costs under control. Traditional methods of operation simply were rendered uneconomic. In consequence a number of families decided to take their money out of the pottery industry at this point.

4. The craft tradition of family firms was not conducive to the employment of scientifically trained managers which the new technology demanded.

5. The increases in output which resulted from mechanisation and improved firing techniques had to be sold at a profit and marketing required expenditures on salesmen, salerooms and promotions beyond the resources of many family firms.

Of course, a few families made the transition and survived and a strong family element remains in a few large public pottery firms. Nevertheless, technological and marketing developments in the 1950s favoured the larger business units and severely restricted the scope of small businesses. The key element in the situation was competition.

Each market was supplied by a number of firms so that prices were determined by market forces. In consequence, firms could not normally increase their prices to offset high costs caused by poor management. Instead poor management resulted in low profits and losses and, in time, the elimination of the unfit. Competition operated effectively by permitting low-cost firms to survive and eliminating the high-cost ones. Some firms, however, succeeded in increasing their prices through manufacturing unique articles for which there were only imperfect substitutes. Examples of this are to be found among the manufacturers of fine china: only Wedgwood produced Jasper* ware and prices may be fixed by the firm rather than by the market; the same could be said of dinner services by Minton or coffee sets by Spode or Royal Worcester. Control over price in these examples is not so much a matter of market imperfections; it is more a matter of enterprise on the part of particular managers. In fact, the firms which have succeeded in controlling costs through mechanisation and long production runs of limited ranges of patterns also tend to be the firms which operate pricing policies as part of their marketing strategies.

The argument rests on the proposition that large production and selling units do not suffer from diseconomies of scale. Why should this be so? The answer is to be found in the nature of management. There are a limited number of able and energetic managers and they tend to operate best in large and growing businesses. Large businesses do not just happen: they are a consequence of the activities of teams of managers. Some small and medium-sized firms can compete successfully with larger firms and we shall see that this is so in the pottery industry; however, it has been the performance of large and growing firms that has been most impressive. The general case for growth by firms has been made as follows:

> Growth is a major motivation. It is a means of securing sufficient size to stabilise commercial results. It allows a company to extend its range of activities and achieve a degree of commercial influence. Growth is also required by shareholders, particularly the ordinary shareholders. A major reason for holding equities or ordinary shares is for capital gain and a hedge against inflation. Over the long term,

---

* Jasper was invented by Josiah Wedgwood, and is described in one of the company's information booklets as follows: ' . . . an unglazed white vitreous fine stoneware which could be stained blue, green, lilac, yellow, maroon or black to provide a suitable background for white classical reliefs or portraits in the same material.' The most famous Jasper piece produced by Wedgwood was the replica of the Portland Vase.

ordinary shareholders expect their shares to increase in value. Unless this expectation is realised at an adequate rate, there may be very serious consequences for lagging companies. Inflation erodes the purchasing power of a given sum of money. Unless capital gains compensate for inflation, and also provide a real, and not an inflationary, reward for holding the securities, the ordinary shares are unlikely to be popular on the Stock Exchange. The end-result may well be take-over. Thus management, particularly in public-quoted companies, has a considerable incentive to ensure that their companies grow.*

In her book *The Theory of the Growth of The Firm†*, Professor Edith T. Penrose presents management, its organisation and its response to challenge, as the main reason why firms grow. Professor Penrose is concerned only with successful firms, that is those with competent and aggressive managers. Her main concern is with their growth rather than their size. The theory is developed as follows. We suppose that a firm has attempted something new, launched a new range of products or transformed production techniques; in due course what was novel becomes routine and the management then finds it has both time and energy to spare. It is Professor Penrose's reasonable contention that growth periodically gives rise to excess capacity in management, and this, in turn, gives rise to a search for new developments and hence further growth. 'Economies of growth are the internal economies available to an individual firm which make expansion profitable in particular directions. They are derived from the unique collection of productive services available to it. . . . At any time the availability of such economies is the result of the process . . . by which unused productive services are continually created within the firm.' She argues that there is 'nothing to prevent the indefinite expansion of firms as time passes'. This is because 'there may be an "optimum" output for each of the firm's product-lines, but not an "optimum" output for the firm as a whole'.‡ Adding new product lines is referred to as 'diversification' and the growth process, from the point of view of an individual firm, may be speeded-up by means of 'acquisitions and mergers'.

Against this background of theory and speculation we can now

---

* F. V. Meyer, D. C. Corner and J. E. S. Parker, *Problems of a Mature Economy*, Macmillan & Co., London, p. 12 (1970).

† Edith T. Penrose, *The Theory of the Growth of the Firm*, Blackwell, Oxford, p. 99 (1959).

‡ Edith T. Penrose, op. cit., pp. 98–99.

look at what has been happening recently in the pottery industry as a direct result of competition. In particular, we will look for concentration of output in the hands of a few firms in each sector of the industry, the growth of firms through takeovers and amalgamations, and the entry into the industry of managers not nurtured mainly on the properties of clays, but trained as accountants, maintenance engineers, designers, personnel and marketing managers and statisticians, chemists and physicists. Such men are dedicated to making radical changes quickly. There were rapid changes in the pottery industry in the 1960s; the mainspring of these changes, we will argue, was the new management teams which had taken-over the commanding heights of the industry.

CONCENTRATION

According to the *Census of Production* for 1963, the five firms with the largest sales accounted for 81% of the sales of the electrical ware sector of the pottery industry, 74% of the tiles sector, 66% of the sanitary ware sector and only 32% of the domestic ware sector. These percentages, particularly the one for domestic ware, must have increased significantly since 1963. The tiles sector is now dominated by one large company formed by an amalgamation between H. & R. Johnson Ltd. and Richard Campbell Tiles Ltd. H. & R. Johnson-Richards Tiles Ltd. and Pilkington Tiles Ltd. now account for at least 80% of output. The high degree of concentration in tiles may be explained by a series of improvements in manufacturing and marketing glazed tiles pioneered by H. & R. Johnson under the firm direction of their Chairman, Derek Johnson, who has a flair for raising and using finance, acquiring businesses and improving and mechanising production methods. One tile firm, Pilkington Tiles Ltd., under its Managing Director Arnold Smith, a research physicist, has held its own in competition with Johnson. The explanation of the high concentration in tiles has been the high degree of standardisation achieved and mass production based on dry dust pressing techniques. The firms which were the first to rationalise production ruthlessly gained a cost advantage which enabled them to take-over or knock out of business the remaining firms in the tiles sector of the industry. Between 1935 and 1970 the number of tiles firms fell from 50 to 14.

Electrical insulators must be perfect, and testing and developing them requires big investments in laboratories and technicians; to achieve adequate throughputs, only a few firms may operate in Britain.

Also, if we ignore the export market, there is only one major buyer—the Central Electricity Generating Board, and this must encourage monopoly on the selling side. Two large producers, Bullers Ltd. and Taylor Tunnicliff & Co. Ltd., merged in 1959 to form Allied Insulators Ltd. Because of the slackening of growth in the British economy after 1966 (the *National Plan** 'to provide the basis for greater economic growth' appeared in September 1965) the demand for high tension insulators fell sharply. This sector of the industry has remained depressed ever since. The third producer of electrical insulators of any size is Doulton & Co. Ltd. which manufactures insulators as well as industrial ceramics, sanitary ware and domestic ware. In this sector of the industry it is difficult to sustain profits because of the buying power of the one major customer.

There is much greater product differentiation in sanitary ware than in tiles; nevertheless, between 1935 and 1970 the number of firms was reduced from 30 to 12. This sector of the industry is dominated by four firms. The largest firm is Armitage Shanks Ltd.; Armitage, which is located near Burton-on-Trent, took over Shanks of Glasgow in 1969, and supplies almost one-third of the market. Ideal-Standard is controlled by American-Standard Inc. and its output is divided between sanitary ware fittings and residential heating appliances; it supplies some 23% of the output of sanitary ware in Britain. Doulton & Co. whose sanitary ware factory is located in Stoke-on-Trent approximately doubled its output by acquiring Johnson & Slater Ltd., which is located near London; the enlarged Company now supplies some 15% of the market. Finally, there is Twyfords Ltd. a firm which only a few years ago was the largest sanitary ware manufacturer; now it supplies some 20% of the market, having refrained from making a takeover. In 1971 Twyfords became part of Reed International Ltd. There is room for four large firms and a few smaller ones in this sector of the industry because of a lack of standardisation; increases in mechanisation may, however, change this. The firms supply builders' merchants and in recent years there has been a substantial amount of consolidation of merchants; big buyers tend to encourage concentration among sellers. Also large sanitary ware factories tend to be more economical to operate than smaller ones; however, because of transport costs, common services are shared between factories and firms do not aim to concentrate their output into one large factory. The number of sanitary ware firms has fallen from 30 in 1935 to 12 in 1970.

* H.M.S.O., Cmnd. 2764.

In the domestic ware sector of the industry too there has been a substantial reduction in the number of firms—from some 230 in 1935 to approximately 75 in 1971. Concentration is less marked in domestic ware than in the other sectors of the industry. This is mainly because producers in the domestic sector supply fashion goods and gift ware, as well as more utilitarian ware, and the extremely wide range of products permits a large number of firms to stay in business. Nevertheless, in the domestic sector, three large firms supply some two-thirds of the market. The three large firms are Wedgwood Ltd., Doulton & Co. Ltd., and Royal Worcester Ltd. Until 1971 there were four large firms; however, in that year Doulton was acquired by S. Pearson & Son Ltd., who in 1964 had formed Allied English Potteries Ltd.—a firm which employed some 5000 persons. The 'Big Three' are considered in detail in Chapter 4. At this point the reader will find it useful if we anticipate consideration of the 'Big Three' by noting the acquisitions made by Wedgwood and Doulton and the formation of Allied English Potteries Ltd.

WEDGWOOD LTD.

Josiah Wedgwood entered into a partnership with Thomas Wieldon of Fenton (Stoke-on-Trent) in 1754 and he founded the firm of Wedgwood in 1759 in Burslem. 'In 1762 Josiah . . . first produced what he described as "a species of earthenware for the table, quite new in appearance, covered with a rich and brilliant glaze, bearing sudden alterations of heat and cold, manufactured with ease and expedition and consequently cheap". Later to be known as Queen's ware by command of Queen Charlotte, this new, inexpensive, and beautiful tableware was, without question, Wedgwood's greatest achievement and contribution to the British pottery industry.'* Not only did Wedgwood make good products, he was able to sell them at a profit. The tradition of good marketing established by Josiah has served the firm well over the years. Under the direction of Mr. Arthur Bryan and Mr. Peter Williams, Wedgwood Ltd. has expanded considerably in recent years by increasing existing capacity and by take-over bids. Unlike Royal Worcester and Doulton, Wedgwood has restricted its activities almost entirely to domestic ware.

Because of its series of takeovers, Josiah Wedgwood & Sons Ltd. is now merely one production unit in Wedgwood Ltd. Wedgwood

* *The Wedgwood Story*, Josiah Wedgwood & Sons Ltd.

extended its range of fine china by acquiring R. H. & S. L. Plant Ltd., Susie Cooper Ltd. and William Adams & Son Ltd. in 1966, and Coalport Ltd. in 1967, then in 1970 a manufacturer of jewellery and a manufacturer of fine glass were acquired. In 1968 the take-over of Johnson Bros. (Hanley) Ltd., the largest producer of earthenware in Britain, marked a radical change in Wedgwood's marketing strategy, as it gained the Group entry to the mass market. Up to this point Wedgwood had catered almost entirely for middle class and, in particular, upper class families. Johnsons were a family firm which had developed over the years a reputation for supplying good quality ware at reasonable prices. In 1970 they acquired a complete breakthrough into the mass market when they acquired J. & G. Meakin Ltd. and W. R., Midwinter Ltd., the two firms having merged in the previous year. Both Meakin and Midwinter had enjoyed reputations for supplying contemporary designs for the home market and traditional designs for export markets at low prices. Both firms before they were taken-over were administered by competent and aggressive managers. In a short space of time Wedgwood had been transformed by means of well-chosen takeovers into one of the world's largest pottery manufacturers. (In 1973 Wedgwood acquired Crown Staffordshire China Co. Ltd. and Mason's Ironstone China Ltd.)

DOULTON & CO. LTD.

In 1815 John Doulton entered into partnership with Martha Jones and John Watts and in 1820 when Mrs. Jones withdrew the assets of the firm, they were valued at £153.* The works were at Lambeth. In about 1827 stoneware pipes were added to the output of stone jars and jugs and domestic utensils. In 1965 Doulton sold its vitrified clay pipe interest to Hepworth Iron Ltd. for £1·75 million. Henry, John's second son, entered the Lambeth works in 1835. The period between 1845 and 1870 was one of steady growth and consolidation. A characteristic of Henry Doulton was the definiteness of his aims. At every stage of his career he knew exactly what he wanted to do, and on this he would concentrate all his powers. Although during his lifetime the firm of Doulton, largely under his guidance and inspiration, grew in scope and ramifications until it embraced almost every branch of the ceramic industry, no new development was undertaken until

---

* Desmond Eyles, *Royal Doulton 1815-1965: The Rise and Expansion of the Royal Doulton Potteries*, Hutchinson, London (1965).

he had satisfied himself that this could be done without affecting adversely those manufactures which were already thriving.'* In 1877 a factory was acquired in Burslem to make domestic ware. Doulton's fine china, excluding acquired companies, is entirely made in the Burslem factory which was reconstructed and considerably enlarged in the 1960s. Doulton continues to manufacture a wide range of industrial ceramics and these activities of the Group are considered in Chapter 9.

In domestic ware Doulton specialised in fine bone china and figurines; by means of takeovers their product range was systematically extended. Entry to the hotel and catering market was obtained by the acquisition of Dunn Bennett & Co. Ltd. in 1968. This long-established firm was a leading supplier of hotel ware for home and overseas markets. Not only did Doulton acquire a complete product range but also new customers and entry into new channels of distribution. In the same year Minton Ltd. was acquired. This firm over the years had been so dedicated to top-quality products that even profits had been neglected; Doulton was admirably placed to sustain quality and inject a profit requirement by means of cost reductions and price increases. John Beswick Ltd. a firm with a well-earned reputation for manufacturing accurate models of horses and dogs and other animals, as well as a wide range of figurines, was acquired in 1969 and with this acquisition the Doulton figurine range was greatly extended. In the same year Webb Corbett, the manufacturer of fine glass was acquired. Thus all the takeovers improved and extended the traditional Doulton product range, a particularly desirable development for a firm which had been establishing its own retail outlets (shops within shops) at home and abroad.

ALLIED ENGLISH POTTERIES LTD.

The Chairman of S. Pearson & Sons Ltd., a large and powerful financial holding company, is Lord Cowdray. The connection between Pearson and the pottery industry has been traced to the marriage of a sister of Lord Cowdray to the London agent of Booths Ltd. Lord Cowdray provided financial assistance to Booths in the 1930s and in 1944 Colclough Ltd., another Stoke-on-Trent pottery, was acquired. Then in 1952 the Lawley Group was acquired. It consisted of a chain

* Desmond Eyles, *Royal Doulton 1815-1965*, Hutchinson, London, p. 54 (1965).

of glass and china shops and the factories of Ridgways and Swinnertons in Stoke-on-Trent which had been acquired in 1945.

In 1964 Pearson decided to augment its interests in pottery and the large and successful Stoke-on-Trent firm of Thos. C. Wild & Sons Ltd. was purchased. Then to provide the new Group with prestige, as well as for other reasons, the Royal Crown Derby Porcelain Co. Ltd. was added. The new Group was in a position to supply a full range of fine bone china and bone china and earthenware. The name Allied English Potteries Ltd. was adopted. It operated 14 factories in Stoke-on-Trent and one in Derby as well as 45 retail outlets and its overseas subsidiaries. Finally in 1971 it was decided by the Pearson Group that further expansion was essential. Doulton & Co. Ltd. was acquired and the constituent companies of Allied English Potteries Ltd. became part of Royal Doulton Tableware Ltd.

In Chapter 4 there is an account of how Allied English Potteries operated between 1964 and 1971. It is a Group which has undergone three major reorganisations in less than 10 years. Its financial strength enabled it not only to compete strongly with Wedgwood, Doulton and Royal Worcester, but permitted it to acquire one of them.

TAKEOVERS AND AMALGAMATIONS

British industry, it is fair to say, has been dominated by takeovers and amalgamations in the past 20 years. In this respect the pottery industry is little different than the rest. Recently, economists have been extremely critical of output being concentrated into the hands of a few large companies. Size, it is claimed, has been pursued for its own sake and for the sake, possibly, of monopoly profits, frequently at the expense of efficiency. This may be so in large measure in many British industries; however, in this respect, the pottery industry is different. In this industry, it would appear, the amalgamations and takeovers have resulted in greater efficiency and an improved industrial structure. A reduction in numbers does not necessarily result in less competition. The formation of the three large groups in the domestic sector of the pottery industry has offered stiff competition to the smaller firms which remain in business. 'Competition between the few in innovating industries is likely to be more intense and effective than competition between the many in technically stagnant industries. Mere numbers cannot ensure useful competition.'*

* B. R. Williams, 'Some Conditions of Useful Competition', *Yorkshire Bulletin of Economic and Social Research*, Vol. 11, No. 2, p. 76 (December 1959).

The Wedgwood and Doulton takeovers were logical and constructive. They dispose of the false argument that takeovers in the pottery industry were mainly a hotch potch of hurried, arbitrary and ill-considered decisions. The formation of Allied English Potteries made good sense and stimulated competition. It will be argued subsequently that the merger of Allied English Potteries and Doulton was beneficial to both companies and to the industry as a whole.

We have noted the takeovers and amalgamations in tiles, sanitary ware and electrical porcelain, and in domestic ware, and how Allied English Potteries Ltd. was the outcome of a series of amalgamations. In Appendix 3 will be found a table of takeovers and amalgamations in the pottery industry since 1948. The table shows that in the domestic sector of the industry 40 parent companies gained control of 105 subsidiary companies: obviously takeovers have played a major part in restructuring the industry.

A takeover requires that if firm X acquires firm Y, firm X should pay more than the value of the assets of Y to the existing shareholders of Y. This implies that the assets must increase in potential value when they move from Y to X. Firm X must expect to provide better management than was provided by the managers of firm Y so as to provide the desired increase in the value of the assets acquired. The pottery industry could be considered to have been ripe for takeovers after 1945 as there existed significant differences in the competence of the managements of firms. Also firms with similar degrees of management competence, by means of amalgamations, could discern benefits to be derived from creating larger management teams.

There have been two main types of takeover in the pottery industry: (1) diversification by firms not in the pottery industry; (2) the rationalisation of existing businesses. Some takeovers were undertaken for tax or retirement reasons, or even for reasons based on friendship or spite, and do not fit into either category. An example of the first type of takeover was the purchase of W. T. Copeland & Sons Ltd. (Spode) by Carborundum Ltd., the American Company which produces grinding materials and equipment. Carborundum decided to diversify its product range and one of Britain's leading fine china manufacturers was a suitable acquisition from their point of view. From Copeland's point of view, the takeover guaranteed their long-run survival independent of (say) Wedgwood, Doulton or Royal Worcester. It also provided them with the capital to finance a major reconstruction of their factory, which was long overdue, and provided the stimulus for expansion in established and new markets. In 1970 the name of the firm was changed to Spode Ltd. In 1964 the Semart

Importing Company, which changed its name in 1966 to Automatic Retailers of America, bought two well-established medium sized firms—Enoch Wedgwood (Tunstall) Ltd., which is an earthenware manufacturer, and Crown Staffordshire China Co. Ltd. Since the acquisition the factories have been reconstructed and modernised, and the two firms together produce a wide range of reasonably priced tableware. Qualcast Ltd., an engineering firm, purchased the Empire Porcelain Company in 1958 and closed it down in 1967 because it failed to yield adequate profits. Few pottery managers appeared to be surprised by the outcome of this particular takeover: 'if they had only asked us, we would have told them that it is easier to make money out of metal than out of clay.' The Great Universal Stores, which sells large quantities of medium-priced pottery, purchased Barratts of Staffordshire in 1948 and Furnivals Ltd. in 1967. Both firms manufacture earthenware. Their output is sold on the market without restriction and Great Universal Stores do not restrict their purchases in any way to their own pottery firms. In 1970 Spode Ltd. made a bid for John Aynsley & Sons Ltd., a well established and respected manufacturer of bone china. The bid was not accepted by the shareholders and instead a bid of some one million pounds was accepted from Waterford Glass. This unexpected takeover makes sense, as top quality glass and china are sold in the same retail outlets. From the point of view of the pottery industry, the great advantage of takeovers by outside firms is the much needed injection of new capital into a sector of the economy where plans for modernisation tend to exceed available financial resources.

The second type of takeover is the rationalisation of existing businesses. The three leading examples in the pottery industry have already been mentioned: Allied English, Doulton and Wedgwood. Another good example of this was the move of eleven firms from various locations in Stoke-on-Trent to an abandoned airport, where new one-storey buildings were erected to house the existing machinery and equipment. The whole operation was supervised by Mr. Bowers, who gave the name Staffordshire Potteries Ltd. to the new Group. Under the direction of Mr. Bowers' two sons Staffordshire Potteries Ltd. has been a most successful company (its bid in 1970 for Wood & Sons (Holdings) did not succeed) and it now dominates the cheap end of the tableware market.

Two aspects of takeovers in the pottery industry deserve to be mentioned. Firstly, firms normally agree to be taken over. An element of compulsion is usually absent. It could be argued that it was an indication of success rather than failure to be taken over by Doulton or Wedgwood. They and the other fine china manufacturers project an image of refinement and superiority that is difficult to resist. It is said that the large firms are frequently approached by smaller ones asking to be absorbed into the larger organisations. The large number of takeovers in 1968 and 1969 may be seen as the large firms taking rapid action to acquire particularly attractive firms before they were acquired by competitors. The acquisition of Coalport Ltd. in 1967 was an example of the initiative coming from the firm acquired and not from the bidder. In 1959 Mr. E. W. Brain of E. Brain & Co. Ltd. acquired Coalport Ltd. because he wanted to keep in production the patterns of a famous firm. His ambition was to preserve something of great value from the past. It is through his efforts that Coalport lines remain in production. Early on Mr. Brain obtained an understanding with Wedgwood that they would be given the first option to purchase Coalport should he wish to sell. In 1970 Coalport was invited to become a member of the Fine China Manufacturers' Association, its output having almost doubled since 1965.

MANAGEMENT TEAMS

The important thing about the new management teams in the pottery industry is that accountants, engineers, etc. supplement the work of managers who have been trained as potters; they have not replaced the traditional skills of the potter—indeed, they have supplemented them. Success in the pottery industry will always depend on the ability of firms to make good pots, however costs may be controlled to greater advantage by substituting a scientific approach for the traditional intuitive hit-or-miss way of doing things. Of equal importance is the ability to sell longer production runs at a profit, and this explains the high value attached to marketing by the new generation of managers. Perhaps 'ordinary men' are more and more replacing the outstanding personalities—the potters—who until recently (and a few still survive) created and dominated the industry. The new style in business has been admirably summarised by J. K. Galbraith. 'The real accomplishment of modern science and technology consists in taking ordinary men, informing them narrowly and deeply and then, through appropriate organisation, arranging to have their knowledge combined with that of other specialised but equally ordinary men. This dispenses with the need for genius. The resulting performance, though less inspiring, is far more predictable.'*

* J. K. Galbraith, *The New Industrial State*, Hamish Hamilton, London, p.62 (1967).

COMPETITION AND MONOPOLY

Firms frequently grow by means of vertical integration, that is to say they produce their own raw materials, machines and components in preference to purchasing them on the market, and they undertake their own wholesaling and retailing. This has not happened in the pottery industry, mainly because it is a small industry and specialist firms supply raw materials, etc. to all the firms in the industry. Firms, however, more and more undertake their own wholesaling, selling direct to retailers, except in sanitary ware and tiles, where sales are made only to merchants. Retailing too remains a task for the specialist and not for the potters. This is because cutlery and glass are normally sold along with pottery, and frequently pottery is sold along with hardware or fancy goods. Shops within shops operated by pottery firms account for only a small fraction of all pottery sales. Nevertheless, firms are attempting to move nearer to consumers by financing and/or operating retail outlets.

The tendency for production to be concentrated more and more in the hands of fewer firms has not, in our opinion, weakened competition. A sufficient number of firms remain in each sector of the industry to provide adequate protection for consumers. Also entry is not difficult; the reason why more firms do not enter the industry is the modest profits that are earned by existing firms, an indication of competition and not monopoly.

The pressures of competition are not restricted to competition between manufacturers of pottery in Britain. British firms meet severe competition in world markets from manufacturers in Germany, Japan, Italy, France and other countries too. The large Japanese and German firms act as a stimulus to British firms to be larger in order to compete more successfully on the world market. In addition, manufacturers face severe competition from products manufactured from non-ceramic materials (plastics, glass, stainless steel, etc.) and from manufacturers of all products which compete for consumers' expenditure, for example, manufacturers of carpets, clothing and motor cars. To the extent that the manufacturers of competing products become more aggressive in their selling it is important that pottery manufacturers retaliate successfully in kind. In many ways competition has increased in intensity, rather than diminished, in recent years.

The structure of the pottery industry was never planned, it 'just growed'. However, its development was controlled by competition, and the pattern that has emerged would appear to be a satisfactory one. It is an industry in which there is no obvious need for government intervention to speed up change or curb monopoly practices.

TRADE AND RESEARCH ASSOCIATIONS

Although pottery proprietors and manufacturers compete fiercely with each other, like footballers and tennis players, most of them remain the best of friends from day to day. Mutual respect, however, rarely leads to genuine and sustained co-operative effort on product development or marketing. Few of the top men in the pottery industry would approve of increased government initiative in industry, at least not in the pottery industry, except perhaps to curtail pottery imports or curb the power of the trade unions. Similar political views contribute towards a community of interests. It is a feature of the industry that the managers of firms which give every indication of being deadly rivals are willing and pleased to assist each other with advice or the loan of equipment or materials in an emergency.

The British Ceramic Manufacturers' Federation (BCMF), which was known until 1962 as the British Pottery Manufacturers' Federation, was founded in 1919. In 1948 the Federation was substantially reorganised. A full-time Director was appointed. One of the main tasks of the Federation is to negotiate rates of pay and conditions of work on behalf of member firms on a regular basis with the Ceramic and Allied Trades Union through the National Joint Council for the Pottery Industry. A number of autonomous Trade Associations or Councils are affiliated to the Federation. They include: General Earthenware (Home and Export) Manufacturers' Association, Staffordshire Potteries Hotel Ware Manufacturers' Association, Fine China and Earthenware Manufacturers' Association, English Bone China Manufacturers' Association, British Ceramic Tile Council Ltd., Council of British Ceramic Sanitaryware Manufacturers, British Electro-Ceramic Manufacturers' Association and the Ornamental Pottery Association.

The Federation, of course, is not an output or price fixing body. A greater degree of co-operation on standardisation, promotion, etc. would appear to have been achieved in industrial ceramics as compared with domestic ware. In domestic ware there are, understandably, conflicts of interest between large and small firms. Competitive forces are essentially destructive and they must be contained in appropriate social structures so that they may operate effectively. The activities of the Federation clearly introduce a degree of co-operation that is essential for the effective working of the industry. It is important to note that the activities of the various trade associations in the Federation play their part, along with market forces, in contributing towards efficiency. Mr Sam H. Jerrett, since 1955, has been a strong and active Director. The consequence has been that on many important issues the pottery industry appears to present a united point of view to foreign governments, government departments, employer and labour organisations and the public at large.

Most of the Associations are in one office block and administrative and clerical staff are shared. A close watch is kept on upward movements in tariffs and quotas abroad, particularly in the United States, and potential changes are opposed. Not all firms are members of BCMF; however, member firms account for nearly 90% of the sales of the industry.

The British Ceramic Research Association (BCRA) with a staff of over 200, is located in Stoke-on-Trent. The Association investigates the properties of raw materials, body systems and making methods and other aspects of pottery manufacture. In addition to its investigations into domestic ware, tiles, and sanitary ware, the Association operates a Refactory Materials Division (furnace linings, etc.), a Heavy Clay Division (bricks and pipes, etc.) and a Special Ceramics Division. Its activities are financed largely by subscriptions paid by member firms, together with a government grant. The Research Association provides pottery firms with high-powered scientists and technologists to help and advise on their technical problems. Although technical advice is frequently obtained by manufacturers from suppliers of fuels, materials, equipment and machinery, and some manufacturers conduct their own research, there is also a great need for a sustained programme of research to be undertaken by a properly constituted and administered research association. The Association also provides testing services for manufacturers. It was established in 1937. The British Refractories Research Association had been established as part of the Department of Scientific and Industrial Research in 1920.*

In contrast to the basic research which is systematically undertaken into the physical and engineering aspects of pottery production, research into the economic, management and sociological aspects of the industry has been neglected.† The BCRA has contributed substantially over the years to improvements in both the health and working conditions of pottery workers.

* *The British Ceramic Research Association Handbook.*
† Their awareness of this has been demonstrated by their appointment of an economist in 1973.

Part II

THE DOMESTIC SECTOR OF THE INDUSTRY

4

THE 'BIG THREE'

The age-old empirical methods to which many British potters paid indiscriminate homage, often mistrusting the findings of ceramic scientists; the dynastic family influence particularly marked in Staffordshire; the lure of quick personal profits without regard for the future needs of their factories, the deeply rooted conservatism and resistance to change in what was still mainly a craft industry; the romantic but often fictitious aura of mystique that pervaded much of the industry—these were insidious dangers in an age in which the fruits of scientific research, modern technology, mechanisation, market research and similar trends in production and selling were being exploited not only by newer industries, untrammelled by the past, but also by modern potters in the United States and other countries.

Desmond Eyles, *Royal Doulton 1815–1965*, Hutchinson, London (1965).

There is no simple answer to the question: why do some firms grow and prosper whilst others stagnate or disappear? An explanation of why some firms succeed would be an amalgam of, among other things, fortunate timing, expanding markets, cost advantages, product developments which provided against competition, availability of capital and avoidance of prolonged labour troubles. The list fails to include what primarily accounted for the growth and survival of a number of tableware firms, namely, enterprising founders and their progeny. Two outstanding examples are John Doulton and Josiah Wedgwood and their descendants.

That many pottery firms lasted only for one, two or three generations is not surprising. In an industry where new entry is easy and the growth of the market slow, it is actually desirable that, regularly, a number of firms disappear. The elimination of a pottery firm could cause little dislocation; new entrants usually purchased existing buildings and equipment and, at least in Stoke-on-Trent, workers might obtain new jobs with neighbouring firms. The time has come, however, to recognise that the great strength of the British pottery industry is concentrated in a small number of large firms and it is mistaken to regard small firms as the typical unit of organisation.

In 1885 the Prince of Wales presented Henry Doulton with the Albert Medal of the Society of Arts. It was presented in the firm's London showroom, and the Choir of Westminster Abbey was present for the occasion. The medal was awarded: 'In recognition of the impulse given by him to the production of artistic pottery in this country. The Council have felt that the establishment of a new industry of this character fully justified the award . . . but, while recording this fact, they wish it to be understood that in making it they also had in view the other services rendered by Mr. Doulton to the cause of technical education, especially the technical education of women; to sanitary science and, though in a lesser degree, to other branches of science.'*

Henry Doulton had agitated along with Chadwick for sanitary reform and he greatly improved glazed pipes, in particular methods of joining them, and he arranged for their production in volume at a time when the market was expanding rapidly. It was his enthusiasm for art ware and his ability to attract outstanding designers to his firm which helped guarantee the subsequent success of Royal Doulton. Once firmly established, tradition can provide a firm with staying-power and even instil into its owners and managers the will to expand.

The Wedgwood family tree (*Figure 4.1*) shows six generations of the Wedgwood family and connections with the firm of Wedgwood. Finally the Hon. Josiah Wedgwood, the fifth Josiah since the founder, retired as Chairman in 1963 and a Managing Director from outside the family was appointed. Josiah Wedgwood is now rightly regarded as one of the truly great British entrepreneurs. He made numerous innovations in the fields of both technology and marketing and he was elected to be a Fellow of the Royal Society in 1783.† The momentum that Josiah gave to his firm only faltered occasionally over the years. Today Wedgwood Ltd. attempts to combine the seemingly

---

*Desmond Eyles, *Royal Doulton 1815–1965* Hutchinson, London, p. 120 (1965).

†N. McKendrick, 'Josiah Wedgwood: An Eighteenth-Century Entrepreneur in Salesmanship and Marketing Techniques', *The Economic History Review*, Second Series, Vol. XII (1959–1960).

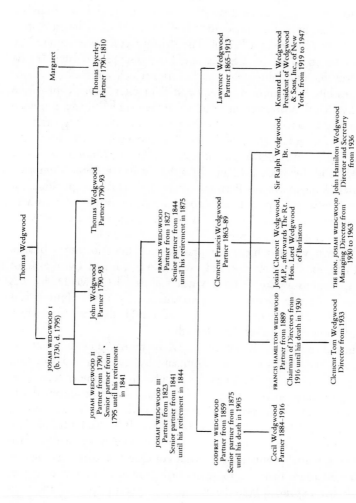

Figure 4.1   The Wedgwood family and the pottery industry

incompatible—tradition and contemporaneity. Josiah V made an important decision in 1936 when he arranged for his firm to leave Etruria, in the centre of Stoke-on-Trent and move instead to a large country estate south of the City. The new site has provided the firm with ample space for expansion in the boom years since 1946.

The third member of the 'Big Three' is Royal Worcester Ltd. which was founded even earlier than Wedgwood in 1751. Three famous factories had been established after 1740 to reproduce Chinese porcelains: Bow, Chelsea and Bristol. The Bristol factory was acquired by Worcester in 1752 and Chelsea and Bow disappeared later. In the long run the only old-established factories which survived outside of Staffordshire were those at Derby and Worcester.* One explanation of why Royal Worcester survived is that the firm has always provided scope for experiment to individual craftsmen and designers.† Its founder, Dr. John Wall, physician, business man and painter, as a potter is obviously not in the same class as Spode, Minton, Wood or Wedgwood. Between 1912 and 1953 the Chairman of the Company, Charles William Dyson Perrins, sustained the firm's reputation for superb craftsmanship and, as a consequence of war-time contracts, diversified into industrial ceramics and electronics. In the 1960s the firm entered into a new period of aggressive expansion.

The firm which temporarily transformed the trio into a quartet is Allied English Potteries Ltd. which was formed by means of amalgamations as late as 1964, when the Royal Crown Derby was acquired. Crown Derby had been founded by 1750. Until it was taken-over, Crown Derby had remained a modestly-sized firm which aimed for survival rather than growth.

The refurbishing of Royal Crown Derby by Allied English Potteries Ltd. and the Coalport China Company by E. Brain and Co. Ltd. and subsequently by Wedgwood Ltd. serve as a reminder that being a long-standing manufacturer of fine china does not automatically guarantee success in terms of profits, growth or influence. W. T. Copeland and Sons Ltd. (Spode Ltd. since 1970) and Minton Ltd. enjoyed the highest reputations for some 200 years, yet neither firm made substantial profits since the death of Queen Victoria and neither permitted more than the most gradual growth of output to occur. Spode has expanded considerably since it became part of the Carborundum group of companies in 1966 and two years later Minton Ltd. became part of Doulton Ltd.

*R. J. Charleston (Ed.), *World Ceramics*, Hamlyn, London, p.254 (1968).

†George Savage, *The Story of Royal Worcester*, Pitkin Pictorials Ltd., London (1968).

WEDGWOOD LTD.

The head office of Wedgwood Ltd. and the factory of Josiah Wedgwood & Sons Ltd. are located in a large country estate five miles from Stoke-on-Trent. The lakes, parkland and trees provide protection against day to day distractions. However, intimate contacts with Stoke-on-Trent have not been lost as large numbers commute each day to Barlaston from the City where Wedgwood was produced for over 150 years.

The isolation contributes to the Company's policy of 'go-it-alone'. It is a company steeped in tradition and immensely proud of its past, yet also a company dedicated to developing new products, techniques, designs, markets and with a vigorous management team. Information

Plate 4.1   Panoramic view of the newly extended Wedgwood hand enamelling department

is fed to Barlaston on changes in the design, pricing, packaging and presentation of rivals products in New York, Toronto, San Francisco, Nassau, Paris, Rome and London. Also information is kept up-to-date on how the firm's own large range of products is selling at home and abroad. Information is more dependable when it is supplied by the firm's own selling organisations abroad and by its own retail outlets at home. The collection and processing of information permits better decisions to be taken than would otherwise be the case on investing in new products or new activities. This is one example of the economies of scale of a large and growing business.

An attractive design studio has been erected at Barlaston to achieve regular close contacts between designers and production and sales managers. The objective is to sustain a product range which is not lacking in flair or distinction. One can discern a strong element of contemporary design in the Wedgwood range in recent years. No longer is it required of trendy persons that they should dismiss Wedgwood designs as safe and old fashioned. Designers must not only know enough about production techniques to avoid imposing impossible demands on men and machines, they must also keep constantly in touch with developments in the fine arts at home and abroad and Wedgwood designers are encouraged to do just that.

Obviously Wedgwood is not isolated in its country retreat. Its contacts are with the wider world that really matters, with fashion, with competitors at the point of sale and with what consumers are thinking and planning. Recently one of Wedgwood's marketing managers made the following confident claim: 'At Wedgwood I do not think we shall be saying anything provocative by stating that design and production excellence are taken for granted and that the success of the company in recent years has been due to the sales orientation of the whole business.' Presumably the possibility of the remark being made 'Josiah would not have been proud of that' about a new Wedgwood product ensures that managers do not compromise with quality standards. It is elementary that successful marketing depends on good products.

So far we have indicated how Wedgwood Ltd. aims to keep in touch with its customers. However, the Company also aims to claim the attention of potential customers and persuade them to spend their money on Wedgwood products. The Wedgwood Group has one substantial advantage over rival firms: the name Wedgwood, which for a long time has been a household word in Britain and America.*
Many customers, particularly in the United States, prefer Wedgwood because of the name on the back of the plate rather than the pattern

---

*Two market surveys revealed that housewives found it difficult to name more than two or three manufacturers of pottery. In 1958, 32% of housewives were not able to mention any. In both surveys Wedgwood was the firm that most readily came to mind; the other firms were Doulton, Crown Derby, Pyrex, Spode and Royal Worcester. (McCann–Erickson 1958, and North Staffordshire College of Technology Department of Management and Business Studies 1967.)

on the front. There are three main ways in which the public are persuaded to favour Wedgwood: by means of sales promotions including advertising, by stores-within-stores, and by the activities of the Wedgwood Public Relations Department.

Sales promotions, of course, must be geared to particular segments of the market and the product range must be such that particularly profitable markets are not neglected. The products of Josiah Wedgwood and Sons were probably purchased by only some 10% of the families in Britain. Except for the odd ashtray or vase received as a present, Wedgwood was used almost exclusively by upper middle class families. Now the Wedgwood Group, which embraces the large earthenware manufacturing firms of Johnson Brothers, Midwinter and J. & G. Meakin, caters for at least 80% of the families in Britain and for most market segments. The Group provides expensive and inexpensive ware, ware for personal use and ware for presents, contemporary and traditional ware, also craft-type ware and hotel ware. In promoting particular lines in magazines and newspapers they are each featured with the Wedgwood name, thereby providing the reader with a reminder about Wedgwood and a glimpse of yet another Wedgwood product. Advertising is not necessarily designed to persuade the reader to purchase the featured product immediately; it may merely be intended to initiate a long process of associating Wedgwood with attractive tableware in the consumer's mind. Advertising to be successful must be sustained and directed towards national markets and large numbers of consumers. When the firms in the pottery industry were nearly all medium or small there was very little advertising and this contributed towards the expenditure on pottery, per household, per year, being pitifully small (£1·25 in 1957). Now the small number of large groups, and Wedgwood in particular, have increased their expenditure on advertising considerably. This must be regarded as a good thing. Its ultimate effect must be to offset advertising by suppliers of competing materials such as plastics, toughened glass, waxed paper and stainless steel.

Glass and china shops provide attractive and frequently efficient retail outlets for pottery manufacturers; unfortunately their high rents and labour costs and an absence of self-service forced many of them out of business. They have been replaced by glass and china departments in department stores and by increased sales by variety stores. Unfortunately, from the point of view of pottery manufacturers, china and glass departments tend to be relegated to attics or basements and their sales staff are not always well-informed about what they are selling. China is bulky, fragile and difficult to display and therefore

Plate 4.2   View of the newly opened 53rd Wedgwood Room at Carmichaels, Hull

it tends to be pushed to one side by more profitable and easier handled merchandise.

The larger pottery manufacturers have reacted positively to this situation and established their own properly operated departments in department stores.* In 1970 there were 51 Wedgwood Rooms in the United Kingdom. The first Wedgwood Room was opened in Marshall and Snelgrove's store in London in 1953. Stores-within-stores had been pioneered by manufacturers of clothing, jewellery and cosmetics, and Wedgwood was the first pottery firm to operate one. Other firms followed and in 1970 Royal Doulton and Royal Worcester operated some 35 retail outlets each. Additional ones are now operated by Poole Pottery and Denby and by other manufacturers. The usual arrangement is for rent for the space to be related to the sales made in the form of an agreed percentage. Until it is sold the stock belongs to Wedgwood who also train and pay the shop assistants. Obviously the stores are expensive to operate and they must absorb a lot of capital; however, they play their part in promoting the firm's products and they permit Wedgwood to sustain direct contacts between their staff and the buying public. An important consideration is that the range of stock offered and the displays are entirely in the control of Wedgwood. In addition, all leading glass and china shops and department stores sell Wedgwood.

Distribution is therefore a planned operation by the manufacturer through its own retail outlets and through selected retailers. Wedgwood dinner sets are not goods that any retailers could expect to obtain from a wholesaler. Perhaps the output of Josiah Wedgwood & Sons before firms were acquired would have justified, say, 25 Wedgwood Rooms, and the acquisition of other fine china manufacturers might have justified operating another 10. Once Johnson Brothers, J. & G. Meakin and Midwinter had been acquired, then a target of 60 Rooms made good sense. With the increased product range the large rooms could offer a really wide range of all qualities and types of tableware, trinkets and ornaments from within the Group. To sustain their attractiveness retail outlets must offer a wide and changing variety of ware. Thus sales promotions, forward integration into wholesaling and retailing and takeovers of firms manufacturing fine china, earthenware and good quality glassware, are seen to be parts of one strategy which involved rapid growth. The third important element in Wedgwood's marketing strategy is public relations.

*Lucien Myers, 'The Trend to Shops Within Shops', *Tableware International* (December 1970).

As the Wedgwood group grew rapidly, so too did their public relations department. A large company cannot afford not to have professionals who maintain close contacts with the press, radio and television and with government departments in Britain and overseas. In a large international company like Wedgwood, which is growing and constantly making changes, there is scope for the public relations officer to find, develop and propagate, items of news. Only a company in which there were just routine happenings could afford to dispense with an active public relations department.

PERSONNEL

A management function which has grown in importance with the expansion of Wedgwood is personnel. This is a branch of management which has been neglected by most pottery firms. Until fairly recently Wedgwood were content to operate training schemes for various operations in the factory; now, however, training is one element in personnel management. In addition to the industrial relations executive, who is responsible for the personnel policy of the group as a whole, personnel managers and officers operate in the various factories. In a large firm, to leave personnel in the hands of production managers is to run the risk of inadequate attention being paid to production planning and control. Also, a personnel department can plan to reduce the incidence of labour turn-over and absenteeism, thereby contributing towards lower costs of production. Whereas personnel managers expect men and women to be awkward at times, not to mention lacking in loyalty and drive, production managers tend to be impatient of human foibles and weaknesses and expect workers to be as dependable as properly maintained machines. The personnel department maintains regular contacts with trade union officials and workers' representatives to maintain good industrial relations at all levels. Of course, not all workers react favourably to the Wedgwood treatment. When we asked one woman pottery worker why she had left Wedgwood, she replied, 'there are too many busybodies at Barlaston.' However, this is not a typical reaction, and Wedgwood would appear to have made a sound investment in engaging a team of professionals to sustain a contented labour force. Wedgwood's personnel and public relations departments combine to maintain regular contacts with school-leavers. Wedgwood's success in recruiting school-leavers contrasts with the failure of many other pottery firms to compete adequately for young workers with firms in other industries.

TAKEOVERS

One explanation given of the Wedgwood takeovers has been that
there was excess capacity in marketing. Presumably the Chairman of
Wedgwood had his own company in mind when he stated in 1963:
'It has been said of our industry that we are production dominated.
My view is that real growth in our industry must come through
forceful marketing of good design and advanced technical achievement;
in consumer goods industries such as ours, the only real progress that
can be achieved must be through marketing.'* Clearly Susie Cooper,

Plate 4.3   An automatic glazing machine at William Adams (Wedgwood Ltd.)

William Adams, Royal Tuscan and Coalport benefited from becoming
part of the Wedgwood marketing organisation at home and, particularly, abroad. Johnson Brothers, J. & G. Meakin and Midwinter,
however, were firms with fairly strong marketing departments when
they were acquired. These three firms were regarded as the three
really go-ahead firms in the middle-range of the earthenware sector of
the industry before the takeovers. All of them had adopted advanced
technology before the takeovers. Wedgwood must have augmented

*Arthur Bryan, 'Pottery to-day', *Journal of the Royal Society of Arts* (July 1963).

and not merely stretched its marketing facilities by its acquisitions in earthenware. Johnson Brothers was a firm that preferred to avoid publicity (we were permitted to interview a director of the firm after guaranteeing that we were not journalists). The firm depended on a reputation admired by its established customers of not 'blowing-its-own-horn', something that could not be said about Wedgwood. All the buyers in department stores that we interviewed stated that Johnson Brothers was their favoured supplier of earthenware. Immediately after Wedgwood acquired Johnson Brothers almost everyone became aware that it was the biggest and best earthenware manufacturer in Britain! A number of non-marketing advantages accrued to Wedgwood as a consequence of its acquisition. Experienced managers joined the Group and junior managers in Wedgwood could be rewarded (tested) with responsible positions in subsidiary companies. An important benefit was the application to the companies acquired of the Group's expertise in accounting and financial control. The outstanding outcome of the takeover, however, has been the transformation of Josiah Wedgwood & Sons Ltd. from a fine china manufacturer supplying a relatively small segment of the market into Wedgwood Ltd., a group with the ability to supply the market from the top to almost the bottom. The acquisition of Johnson Brothers was a completely unexpected move for a manufacturer of fine china to make. It could be compared with an amalgamation between the Royal Ballet and Mecca Dancing, or Marks & Spencer and Woolworth.

The expansion of Wedgwood in the 1960s was not merely a matter of acquisitions; in addition, production capacity at Barlaston was increased and oven-to-tableware was produced for the first time in quantity. Stoke-on-Trent firms had traditionally specialised in bone china and earthenware, leaving stoneware to be manufactured elsewhere. Production and sales of bone china dinner sets, the backbone of the business, was increased substantially in the 1960s. When added together, internal expansion and takeovers provided the Group with extremely rapid growth and most important, substantial increases in exports. Production methods at Barlaston improved after members of the Johnson family spotted weaknesses and set about eliminating them. Late in 1973, Wedgwood acquired two more firms: Crown Staffordshire China Co. Ltd. and Mason's Ironstone China Ltd.

The management organisation chart (*Figure 4.2*) shows how the Wedgwood Group has been split into fine china and earthenware divisions with five categories of management services being made available to both divisions and overseas companies. It is significant that each of the management services have direct access to the parent company board.

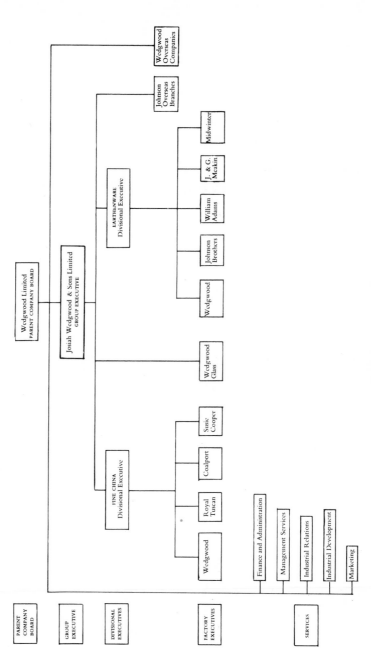

Figure 4.2 Organisation of the Wedgwood Group

Table 4.1 WEDGWOOD LIMITED AND SUBSIDIARIES

United Kingdom	Josiah Wedgwood & Sons Ltd. operating through the following divisions: Wedgwood William Adams Royal Tuscan Susie Cooper Coalport Johnson Brothers Wedgwood Glass J. & G. Meakin Midwinter Abrahams Wedgwood Rooms Ltd. Johnson Brothers (Hanley) Ltd. Coalport China Limited
United States	Josiah Wedgwood & Sons Inc. of America
Canada	Josiah Wedgwood & Sons (Canada) Ltd. Coalport China (Canada) Ltd. Jackson's Chinaware Ltd. Johnson Brothers Hanley (Canada) Ltd. Carleton Securities Ltd.
Australia	Josiah Wedgwood & Sons (Australia) Pty. Ltd.

Year ended 27 March 1971

Turnover £16 973 000
(1970: £16 530 000)
Profit before taxation
£1 334 000
Total capital employed
£6 616 647

Arising on Trading	(%)
Gross trading income	100
Wages and salaries	50
Materials	9
Fuel and power	4
Maintenance	2
Selling and distribution	14
Administration and general expenses	6
Welfare	4
Taxation	4
Dividends to shareholders	2
	95
Available for re-investment in the business	5

Profit Trends	Year	(£'000)	Profits as % of capital employed
	1961	370	
	1962	397	
	1963	292	
	1964	422	
	1965	514	
	1966	640	22·9
	1967	839	23·1
	1968	1 394	19·4
	1969	1 798	16·7
	1970	1 605	16·7
	1971	1 617	12·0
	1972	1 761	9·4

Sources: Wedgwood *Annual Report* (1971) and Moodies Services Ltd.

DOULTON & CO. LTD.

In 1971 Doulton & Co. Ltd. was acquired by S. Pearson & Sons Ltd. The Pearson Group had established Allied English Potteries Ltd. in 1964. After the takeover it was decided that the Allied English factories and showrooms and retail outlets should be absorbed into the Doulton Group whilst retaining their various trading names. Also it was decided the Doulton industrial ceramics factories and research laboratories should be integrated into the engineering division of the Pearson Group. (The Doulton Group, however, continues to include sanitary ware, insulators and industrial products.) The reorganisation following upon the 1971 takeover will take many years to complete so that, writing in 1973, only tentative statements about the consequences of the takeover can be made. In 1972 Royal Doulton Tableware Ltd. was formed to include Doulton Fine China, other Doulton tableware companies and the various companies of what had been Allied English Potteries Ltd. In what follows, Doulton & Co. Ltd. and Allied English Potteries Ltd. will be described as they operated up until the time of the amalgamation. A final section will consider some of the implications of the amalgamation that has produced Britain's largest tableware manufacturer.

In the early 1860s Henry Doulton decided to supplement the production of sanitary ware and stone jars, etc. in his Lambeth works with some decorated stoneware and earthenware pieces. As a consequence of interesting designs and good workmanship the new venture was successful. In 1877 a pottery was acquired at Burslem (Stoke-on-Trent) to make tableware and ornamental ware. Finally, in 1884 bone china bodies were introduced. It is recorded that at the 1910 Annual General Meeting, Lewis Doulton pointed out that 'although the name of the Company was chiefly associated in the mind of the public with the Lambeth and Burslem art-wares, it was mainly on the utilitarian drain-pipes and sanitary wares that the financial success of the whole enterprise depended'. By 1969 this state of affairs had changed considerably: 65% of the profits before tax were accounted for by fine china and industrial products accounted for 63% of the turnover. As may be seen from *Table 4.2* the percentages in 1970 were 72% and 52%. In Chapter 9 there is an account of the activities of Doulton in industrial ceramics and reasons are given for profits falling sharply in the 1960s.

The various activities of Doulton are widely scattered geographically as a consequence of decisions made during a period of over 100 years. The firm was established in Lambeth. Subsequently domestic and ornamental ware production was transferred to Burslem, Stoke-on-Trent, and later sanitary ware production was transferred from London to Tamworth in 1935 and subsequently industrial ceramics were hived-off and moved to Stone; both Tamworth and Stone are in Staffordshire. The Company's research laboratories were situated at Chertsey in Surrey. The distances between Head Office in London and the various factories suggests that for long periods of time the factories operated independently of each other. In day-to-day operations there was seldom any need for contact between, say, the electrical ware and fine china managers, or the industrial ceramics and the sanitary ware managers. However, each specialist concern would have contacts with the Research Laboratories relating to particular projects and problems. It was the task of Head Office to co-ordinate activities by means of financial controls. In addition, Head Office supplied public relations facilities to all of the Company's Divisions, and management recruitment and training was conducted on a Company-wide basis. Doulton tends to be much more science orientated than is usual in pottery firms. Their production managers were sometimes recruited from their research staff. Over the years at Doulton it was found that research proved to be an excellent background training for general management positions. Marketing was not neglected and the Company grew and flourished on a basis of providing a wide range of new and improved products, from wash hand-basins through cream jugs to insulators and moulds for turbine blades.

In the Research Laboratories, high-calibre chemists, physicists and engineers investigated the possibilities of new processes and materials. In addition, research and development and testing work was conducted at Tamworth (electrical ware) and Stone (industrial ceramics). Although the research work was devoted mainly to industrial ceramics, the fine china division had access to research personnel who would not normally be available if all they had to do was investigate the more conventional ceramic materials and manufacturing processes. At Chertsey, making machines and kilns suitable for fine china production were designed and developed. It was the firm's belief that adequate equipment could not always be obtained from outside suppliers. This organisation made good sense so long as the markets for industrial ceramics were booming. Once the prolonged recession in building activity became established in the 1960s the overhead costs of the research expenditures were no longer a spin-off for fine china; instead, research activity became a claim, and burden, on the Company as a whole. In 1970 research activities were severely curtailed. (At the time similar severe cuts in research staff were made in many industries in Britain and in other countries too.)

In the late 1950s a Doulton management team was given the task of perfecting a new tough translucent porcelain body. (Given that bone china is regarded by British pottery manufacturers as the last word in tableware bodies, it was a radical task for a fine china manufacturer to undertake.) Doulton wanted to manufacture ware that would stand up to comparison with bone china yet could be sold at considerably lower prices. The research was undertaken after it had been discovered that there was a large potential market for a new ceramic body. Market research had revealed the possibilities of market penetration available to an established firm offering ware at reduced prices. It appeared that sales would not be at the expense of fine bone china, instead it would affect mainly the manufacturers of high-priced earthenware and Doulton intended to withdraw entirely from that line of production. Rival firms claim that the Doulton research team merely introduced continental hard porcelain into Britain by substituting feldspar for bone, their main contribution being a new name for an old product—English Translucent China (ETC). However, much more was involved; for example, a major achievement was to make ETC using equipment and firing temperatures designed for bone china. Workers from the Burslem factory report that for a long period they produced ETC believing that it was bone china.

ETC was introduced in 1960. The new body was only one essential ingredient of a major marketing operation. Take prices for example. ETC sells at prices which are approximately half the prices asked for fine bone china. Some saving in costs is achieved by substituting a cheaper raw material for bone, but the major saving in cost is the result of really long production runs. ETC meant the introduction of a night shift and the substitution of men for women workers; also the warehousing of large quantities of white ware so that decorating could be a continuous process. The increased volume of sales for each pattern permitted a lower mark-up to be made than is normal for fine china. The design department of Doulton developed new shapes and patterns for the new material, so it was adequately promoted at home and overseas. ETC has been referred to as 'a major event in ceramic technology'; it was also a major marketing event.

Doulton offered price-cuts in bone china in addition to the low-priced ETC. These are based on a limited number of new shapes and patterns modestly decorated and produced in large quantities. Doulton entered the 1970s with a product mix strategically designed for profit and growth. The product range was completed by a wide range of traditional and new bone china shapes and patterns supplemented by Doulton figurines, *rose flambé* ware and toby and character jugs, all well established products.

The development of ETC was only one part of Doulton's planned expansion in the 1960s. In addition, plans were laid down in 1966 to rebuild stage by stage and expand the whole production complex at Burslem and the completion of this undertaking at a cost of some £1¼ million has resulted in a substantial expansion of output. Doulton has also expanded by means of a series of takeovers. As argued in Chapter 3, the Doulton takeovers extended and supplemented existing activities and in no sense were takeovers merely for the sake of acquisitions. In particular, they fitted in well with the policy of establishing Doulton Rooms. The rooms are operated on similar lines to the Wedgwood Rooms, which have been described, and they permitted Doulton to move as close to consumers as possible. Developing new products and markets, rebuilding the main factory and coping with a series of takeovers has kept the Doulton Fine China management team at full stretch—a state of affairs which is essential for a management team to progress from strength to strength. In big-business the more one undertakes the more experience is built-up and the more capable does the management team become of undertaking further expansion in the future.

Whereas Wedgwood had moved into earthenware on a substantial scale and on a permanent basis, Doulton in contrast had abandoned earthenware bodies entirely. From the point of view of a ceramist, earthenware results in a poor body. It is porous, and much weaker than bone china or porcelain, it cannot stand-up to the same stresses of heat and cold and it is more liable to craze. However, it is relatively easy to decorate, has a pleasing and unique texture, and most important —at least in Britain—consumers like it. The divergencies of policy suggest that in Doulton the scientists have more to say about the

Table 4.2 DOULTON AND CO. LTD.: TABLEWARE DIVISION

Royal Doulton Tableware Ltd. Management Company formed in 1968 to co-ordinate the activities of all the fine china companies in the Group.

Doulton Fine China Ltd. Formed in 1956 to continue and develop at Burslem, Stoke-on-Trent, the manufacture of fine china tableware figurines, animal models, *rouge flambe* and other decorated ceramics.

Doulton & Co. Inc. (New York); *Doulton & Co. (South Africa) Pty. Ltd.* (Johannesburg); *Doulton & Co. (Canada) Ltd.* (Toronto); and *Doulton Tableware Pty. Ltd.* (Sydney). (Four companies responsible for the sale of Doulton fine china products.)

Minton Ltd. Acquired in 1968. The famous firm founded in Stoke-on-Trent in 1793. Also Minton Inc. (New York) and Minton China of Canada Ltd. Companies which sell Minton products.

Dunn, Bennett & Co. Ltd. Acquired in 1968. Long established makers of vitrified and restaurant tableware.

John Beswick Ltd. Acquired in 1969. Manufactures popular animal models, figurines and other decorated ceramics. Founded in 1890.

Webb Corbett Ltd. Acquired in 1969. Established in 1897 and producer of high quality hand-crafted full lead crystal glass tableware and ornamental glass.

Year ended 31 December 1970

	Turnover (£'000s)		Profit before tax	
	Tableware	Industrial	Tableware	Industrial
	9 064 (48%)	9 734 (52%)	1 194 (72%)	459 (28%)
Entire Group				

Total capital employed £16 052 000
Employees (1968) 4 130

Profit trends

Year	(£'000s)	Profits as % of capital employed
1966	1 156	16·0
1967	1 318	15·4
1968	1 733	13·6
1969	1 660	11·6
1970	1 543	9·6

Sources: Doulton & Co. Ltd. 'A Pattern of Progress' and Moodies Services Ltd.

body that should be marketed than is the case in Wedgwood. Presumably the Doulton managers would argue that to sell relatively low-priced earthenware could damage their fine china image—let Wedgwood take the risk and good luck to them. (Doulton and Wedgwood are more inclined to admire each other than run each other down.) The 1971 takeover, however, completely changed the product strategy of Doulton. The large modern earthenware factories operated by Allied English Potteries Ltd. (and their medium-priced bone china factories, too) rendered the combined Doulton/Allied English product range similar to that offered by Wedgwood, Johnson Brothers and J. & G. Meakin Ltd. In a sense the merger compensated (perhaps over-compensated) for Doulton's inability to penetrate down the market to the bottom where millions of families are involved, families which could not afford or did not feel inclined to purchase English Translucent China.

ALLIED ENGLISH POTTERIES LTD.

This company was formed in 1964 with the merger between the Lawley Group Ltd. and Thomas C. Wild & Sons Ltd. The name Allied English Potteries Ltd. (AEP) was chosen, perhaps, for its neutrality: it lacks the arresting quality of the names Worcester, Wedgwood and Doulton. The Group could have adopted the well-known name of the Royal Crown Derby Porcelain Co. Ltd. which had been acquired in 1964. However, that would have meant a similar marketing strategy to that adopted by the other three large groups: emphasising the Group name along with, or at the expense of, the names of member firms. Instead, AEP decided to be different; by not emphasising the Group they would encourage associations of member firms to project their own identities and brand names. This strategy was there from the beginning but it was not made crystal clear to all concerned until 1969, when a complete reorganisation of the Group was undertaken.*

The first five years of the new Group's existence must have given rise to considerable internal tensions and misunderstandings; however, a good profit record swamped the discontents. Individually the larger firms in the group did very well and in the years 1965, 1966 and 1967 they were admirably placed to satisfy a world-wide shortage of competently made and reasonably priced tea and coffee sets. In 1968

*'Allied English Potteries' Reorganisation into Self-contained Units', *Pottery Gazette and Glass Trade Review* (June 1969).

and 1969 the Group's rate of profit fell considerably and a need for radical changes became apparent. Member firms had been successful family businesses and most of them were still being managed by members of the original families. Thus for a few highly successful years in the beginning the Group was little more than a loose collection of family businesses. It was large, yet it seemed to lack the cohesion necessary to compete on equal terms with the other members of the 'Big Four'. In particular, the parent company was not represented by a chief administrative officer in Stoke-on-Trent; thus Head Office appeared to be unduly remote and not properly informed about day-to-day matters. We will not dwell here on the early years; instead, we will consider the organisation and performance of the Group after Mr K. M. D. Mills was appointed Chairman. The Chairman was in charge of day-to-day operations of the Group in Stoke-on-Trent and he directly represented the Spearshaft Industrial Group Ltd., which was the controlling company. From the point of view of the pottery industry the Chairman was an outsider. We have argued, however, that the pottery industry has benefited in recent years from managers entering from outside and developing around them a management team which includes experienced potters. AEP is a good example of this type of development.

In 1969 AEP was transformed from a collection of family businesses into an integrated commercial organisation. Within AEP seven main companies were formed, each with its own management team and general manager. Each was charged with earning a stipulated return on the capital provided, taking one year with another. New capital was allocated between the member companies according to their past performances and the expected rates of return over costs of investment projects. Because each of the companies supplies particular markets, they do not compete directly with each other. Each company complemented the other and wasteful duplication was not a major problem. Thus each company has its own production units and its own showrooms and sales staff (in addition AEP operate showrooms for all its companies). The companies would introduce new lines, withdraw old ones and price and merchandise as they thought fit. In fact, the only serious restriction on the activities of the member firms was their need to earn profits consistently. The public is constantly reminded of the products of each company and here AEP is inclined to keep in the background. However, from the point of view of member firms, AEP was very much a force to be reckoned with and a source of key management services. The logic of the organisation was that each member firm was strengthened from the centre and per-

mitted ample scope for independent action, subject, however, to
strict financial control. It was justified if the firms were weaker operating outside the Group than in it. If, in other words, there were genuine
economies to be gained by operating on a substantially larger scale.
Size, of course, does not necessarily give rise to lower costs or increased
revenues (the Co-operative Movement has illustrated that). It all
depends on the choice of products, processes and divisions and the
extent to which a coherent and effective administration is achieved.
Three benefits obtained from the formation of the large organisation
may be noted:

1. The ups and downs of family businesses which tend to be
   directly related to the age, interests and energies of members of
   the controlling families were eliminated. The more impersonal
   type of organisation may be dull and bureaucratic; nevertheless,
   it tends to be more consistent in its operations from year to year.
2. In marketing it is desirable to offer a wide range of products in
   a wide range of materials and styles. In a small firm this prevents
   production economies from being realised; in the large firm
   variety does not prevent low production costs from being
   realised.
3. Small firms operate more or less anonymously: they are known
   only to their regular customers and suppliers. The large organisation uses its public relations and marketing departments to
   render its brand names familiar all over the world.

An important consequence of large-scale operation is likely to be
that the goodwill generated with managers, labour force, wholesalers,
importers, retailers and consumers will contribute towards overall
improvements in efficiency. Goodwill takes a long time to develop.
Doulton and Wedgwood kept at it for 100 and 200 years; AEP could
not be expected to have succeeded in less than a decade.

The firms included in the original amalgamations and takeovers
enabled AEP to provide a comprehensive range of products without
one division of the Group competing unduly with any of the others.
(The domestic sector of the pottery industry is composed of groups
of firms which compete directly with each other; in contrast, competition between the various groups or sectors of the industry is not
severe.) In 1972 the seven production divisions of AEP were as follows:

THE ROYAL CROWN DERBY PORCELAIN CO. LTD.

Since Royal Crown Derby was acquired in 1964 the factory has been extensively modernised. The elaborately decorated and shaped bone china tableware, figurines and ornaments which require hand craftsmanship, continue to be produced. The uniqueness of the wares have prevented high prices and price increases from curtailing demand. The Royal Crown Derby name has provided AEP with a valuable asset. Even consumers who are not attracted by Crown Derby find it difficult not to be impressed by the associations of quality and dignity of a venerable name. All the cheaper products of the Group inevitably benefit from being associated with the Group's prestige member.

PARAGON BONE CHINA LTD. AND ROYAL ALBERT LTD.

Formerly Thomas C. Wild & Sons Ltd., these two units produce large quantities of better quality traditional dinner, tea and coffee sets in their five factories.

RIDGWAY POTTERIES LTD.

This division includes earthenware factories making traditional and contemporary designs and oven-to-tableware. Also a large bone china factory geared to mass-producing a few standard shapes and decorations. The Ridgway Potteries provide cheaper wares than do Paragon and Royal Albert.

RIDGWAY HOTELWARE

Hotel ware tends to be heavier, more robust and less highly-decorated than domestic ware. AEP operate two factories which specialise in hotel ware. It is sold under the trade name 'Steelite' (AEP Catering Supplies Ltd. acts as a wholesaler of glass and hotel ware, including as much 'Steelite' as is reasonable) to the catering industry.

RIDGWAY ROYAL ADDERLEY FLORAL

Produces hand-made bone china floral posies and figurines which are not usually as expensive as those manufactured by Crown Derby.

In addition to the seven production units and AEP Catering Supplies Ltd., Lawleys Ltd. is a chain of 45 china and glass shops which sell a full range of British and imported domestic and ornamental wares. The Lawley shops provide AEP with useful direct contacts with consumers. AEP also operates overseas distribution companies in Australia, Canada, the United States and Belgium.

Each division of AEP, except for the Lawley shops, was a conventional and familiar collection of factories and showrooms for manufacturing and selling pottery. The more interesting part of AEP, from our point of view, was the Group Administrative Centre. Four directors were each responsible for an area of the Group's activities: production, administration, marketing and finance. All the Group's factories benefited from design, public relations, engineering and research services provided on a group basis. Also the personnel officer in each division operated appointment and training schemes which were co-ordinated by the Group's personnel manager. Although the key principle in the organisation of AEP was decentralisation, nevertheless this did not prevent full advantage being taken of services that would be uneconomic to provide in smaller business units. By means of regular meetings between the directors and company managers, sources of inefficiency are kept under review. Properly conducted, regular meetings introduce an important educational element into the Group. In turn this contributes towards augmenting the effectiveness of the management team.

The contrast between the organisation of AEP and Doulton, Wedgwood and Royal Worcester was striking. AEP was decentralised whereas the other three firms were and are highly centralised. Wedgwood and Doulton acquired firms not merely to provide them with management services: the objective was to improve management and integrate them into the Group. The merger between AEP and Doulton therefore presents a series of difficult, though fascinating, problems for the new management team. (The fact that AEP consisted of fairly self-contained production units at the time of the amalgamation possibly renders it less difficult than it otherwise would be to transfer control and co-ordination to Doulton Fine China.) Much of the credit for the success of Doulton Fine China belongs to the firm direction provided by Mr. R. J. Bailey; he would appear to be ideally qualified to take advantage of the opportunities inherent in the amalgamation. AEP offers Doulton a number of modern and well-equipped factories and scope for expansion among other things. Their strength in hotel ware will supplement that of Doulton which exists because Doulton acquired Dunn Bennett & Co. Ltd., then the leading manufacturer of hotel ware, in 1968. The Lawley shops offer Doulton a welcome addition to their shops-within-shops, while the increase in the product range which AEP bring to Doulton must benefit their retail operations. From the point of view of AEP, the takeover means that a period of consolidation gives way to a new period of reconstruction and growth. In Chapter 3 it was argued that most of the amalgamations and takeovers in the pottery industry were constructive and have improved

Table 4.3 ALLIED ENGLISH POTTERIES LTD.

A subsidiary of Spearshaft Industrial Group Ltd., the ultimate holding company is S. Pearson & Sons Ltd. The various companies now form part of Royal Doulton Tableware Ltd.

Operating subsidiaries during 1970
A.E.P. Catering Supplies Ltd.
The British American Glass Co. Ltd. (Registered in Scotland)
Lawleys Ltd.
Paragon China Ltd.
Rhodes & Watson Ltd. (Registered in Scotland)
Ridgway Potteries Ltd.
Royal Albert Ltd. (formerly Thomas C. Wild & Sons Ltd.)
The Royal Crown Derby Porcelain Co. Ltd.
Shelley Furnaces Ltd.

Allied English Potteries Inc. (Incorporated in the United States, November 1970)
Allied English Potteries Europe S.A. (Incorporated in Belgium)
Allied English Potteries Pty. Ltd. (Incorporated in Australia)
Ridgway Potteries Canada Ltd. (Incorporated in Canada)
Royal Crown Derby Canada Ltd. (Incorporated in Canada)
All subsidiaries are wholly owned with the exception of Allied English Potteries Inc. (51%).

Year ended 31 December 1970

Sales to external customers	£10 712 000
Profit before taxation	£797 000
Capital employed	£6 271 000
Number of employees	5 000

Profit trends

Year	(£'000s)	Profits as % of capital employed
1965	1 178	19·7
1966	1 036	16·4
1967	897	14·1
1968	676	10·8
1969	739	11·6
1970	797	12·1
1971	712	10·8

Sources: Allied English Potteries Ltd. *Report and Accounts* (1970) and Moodies Services Ltd.

the structure of the industry. There is good reason to believe that the AEP/Doulton merger will revitalise both participants.

Presumably the Pearson Group expects to earn a comparable rate of return, over the long-period, on capital investments in its wide range of activities. This is achieved by adjusting the amount of capital in each activity to achieve, roughly, equal marginal returns. Presumably if the return drops significantly below the desired level for a number of years, or if it is expected to drop, then the Group would remove its capital entirely from that sector of the economy and transfer it elsewhere. Between 1968 and 1971 the profit rate of AEP varied between 11% and 12% so that pottery must have been one of Pearson's investments that required to be justified from time to time. Unless some radical changes were made it must have been difficult for the Board to be satisfied that eventually the rate of profit would be raised permanently to a higher level. The acquisition of Doulton and the substitution of a centralised for a decentralised administration, combined with the elimination of competition between the two Groups, rendered it likely that higher profit rates would eventually be achieved and sustained.

The requirement that each division of AEP should achieve profit rate targets must have raised additional problems. It could in practice lead to unduly high prices in highly competitive markets or excessive expenditure on selling costs to prevent sales from declining. The aim to supply all sectors of the market (provide full market coverage) could conflict with the need to reduce activities which were insufficiently profitable. In the event the merger with Doulton presents the new management team with the task of devising a programme of long-term rationalisation. It is this aspect of the merger which suggests that it was well-conceived and that it should lead to substantial changes and improvements.

THE ROYAL WORCESTER PORCELAIN COMPANY LTD.

The Royal Worcester Porcelain Company Ltd. is an operating subsidiary of Royal Worcester Ltd. and was established as early as 1751 by Dr. John Wall. At Worcester the old factory concentrates on the production of bone china and figurines, and the new factory, which was opened in 1970, on porcelain, in particular, decorated oven-to-tableware. Since the end of the 1950s, a new management team, composed of outsiders working alongside managers with many years of service with the company, and blending traditional working methods with modern techniques, adopted more adventurous and aggressive policies. The Managing Director from 1964 to 1970 was W. B. Dunn from Jaguar Cars. New product lines were developed for specific markets which were identified by market research. The product range, with the aid of a computer, was rationalised to improve profit performance. A sustained programme of mechanisation was undertaken. However, because of the growth of sales, the labour force expanded. During the 1960s production and marketing techniques were comprehensively reviewed and a number of improvements were introduced.

Royal Worcester specialises in the production of hard porcelain which is not produced by any other tableware producer in Britain. (However, Doulton's ETC is, presumably, a modified type of hard porcelain.) It is aimed at the expanding high-class catering market as well as households. In addition, the body is suitable for laboratory ware. The hard porcelain is cheaper than bone china because there is some saving in the cost of raw materials. More important, however, is the fact that it is easier to manufacture and firing losses tend to be lower than is usual for bone china. It is translucent, it can be decorated, it is tough and it is particularly difficult to chip.

Royal Worcester's hard porcelain is designed for thermal shock resistance and mechanical strength which are essential for flame-proof cooking ware and laboratory and hotel ware. Traditional hard porcelain is more brittle and chips more easily than bone china, these shortcomings have been overcome in the Worcester porcelain with the result that it is less translucent than Continental porcelain. The lack of interest in hard porcelain shown by British tableware manufacturers may be explained by it having been regarded as unduly brittle and inferior to bone china. Bone china is regarded as a superior body because of its subtle texture and the range of colours that can be used for decoration. Nevertheless, bone china is more expensive to manufacture and it is less resistant to heat and cooling shocks than is hard porcelain. Also it is less hard and scratch resistant than with hard porcelain. However, the production of bone china in large quantities has not been neglected by Royal Worcester—it is only earthenware that has been neglected. In 1958 Royal Worcester acquired Palissy Pottery Ltd. in Stoke-on-Trent, a producer of high quality earthenware. (Bernard Palissy, 1510–1590, master-potter and creator of new shapes and glazes in earthenware.) By owning Palissy Pottery, Royal Worcester can offer a wide range of tableware in the 36 Royal Worcester Rooms in Great Britain and in the few rooms operated abroad.

During the period in which both Doulton and Wedgwood acquired firm after firm, Royal Worcester avoided making acquisitions. Part of the explanation of a lack of action in the takeover field may lie in the firm's substantial commitments in industrial ceramics and electronics. Also the extremely wide range of tableware and ornamental ware

Table 4.4 ROYAL WORCESTER LTD.

Operating subsidiaries during 1969 were:

Ceramics Division	The Worcester Royal Porcelain Company Ltd.
	Royal Worcester Industrial Ceramics Ltd.
	Palissy Pottery Ltd.
United States	Royal Worcester Porcelain Company Inc.
	Worcester Silver Company Inc.
	Royal Worcester Industrial Ceramics Inc.
Canada	Worcester Royal Porcelain Company (Canada) Ltd.

Electronics Division	Welwyn Electric Ltd.
	Welwyn Electric (Estates) Ltd.
	Strainstall Ltd.

In addition, 3 companies in the United States, 2 in Canada and one in Australia.

An Associated Company is Worcester House A/S (Copenhagen).

Year ended 31 December 1969

Activity	Turnover ($£'000$)	Profit before tax ($£'000$)	Total profit (%)
Ceramics	4 882	655	57
Electronics and Engineering	6 870	489	43
	11 752	1 144	100

The average number of persons employed (including Directors) by the Group in the United Kingdom was 4 307.

Profit trends (Group as a whole)

Year	($£'000$)	Profits as % of capital employed
1965	573	17·1
1966	603	14·9
1967	890	16·0
1968	1 288	20·8
1969	1 464	22·0
1970	1 228	14·1

Sources: Royal Worcester Ltd., *Report and Accounts* (1969) and Moodies Services Ltd.

manufactured at Worcester may have stifled the desire to extend the product range by means of acquisitions. At present the Company manufactures some 6000 items from over 100 different lines.

5

DOMESTIC WARE
MANUFACTURERS

The domestic sector of the British pottery industry may be regarded as divided between two groups of firms: (1) the 'Big Three' groups: Pearson (Allied English/Doulton), Wedgwood and Royal Worcester; and (2) smaller firms and groups which, at least at the time of writing, are independent of the 'Big Three'. The firms in group (1) were considered in the previous chapter. In this chapter our concern is with the firms in group (2).

In Chapter 4 all the firms concerned were considered, there being only four of them. A similar procedure, if attempted in this chapter, would only lead to a tedious listing of names, products and descriptions, and an undue amount of duplication. Many of the smaller firms manufacture basically the same products using fairly standardised machines, kilns and equipment, and they pursue similar policies. In spite of similarities, however, there are significant differences between firms, so that the firms described in this chapter cannot be regarded as representative of all the firms in each division of the domestic sector of the industry. It is preferable to consider particular firms rather than attempt to generalise about small, medium and large ones. The differences have resulted from the various ways firms have survived the ravages of wars and depressions. The firms which survived developed goodwill with their distributors and their workforces and characteristics which transcend each new generation of proprietors and managers. Also at each point of time particular men and women impose their personalities on the firms they control. The top men in the industry have not been inclined to be easy-going; they enjoyed having things their way.

The firms we have selected are both distinctive and representative. One important limitation is that they are representative only of successful firms. Nearly all the surviving firms have something distinctive to offer. Perhaps it may be traditional designs or trendy designs, or specialised wares for hospitals or ships, or particularly cheap ware, or expensive wares which may be purchased as an investment. As we shall see, all the firms included in this chapter specialise in particular groups of products or in particular markets with large product groups.

The firms have been selected to provide examples of activity in the various sub-sectors of the domestic sector of the industry. The fine china sector is represented by Spode Ltd., Minton Ltd., and Mason's Ironstone China Ltd. Alfred Clough Ltd. and Staffordshire Potteries Ltd. between them account for most of the opposite end of the market —low-priced earthenware. Alfred Clough Ltd. also manufactures more expensive tableware. J. & G. Meakin Ltd. (now a subsidiary of Wedgwood Ltd.) and T. G. Green Ltd. are earthenware manufacturers who supply the middle range of the market, as does Thomas Poole and Gladstone Ltd. which also supplies bone china tea and coffee ware. Three firms, Arthur Wood and Son (Longport) Ltd., Elijah Cotton Ltd. and Portmerion Potteries Ltd., all supply some earthenware tableware; however, the bulk of their output consists of ornamental ware. (George Wade & Son Ltd. also produces ornamental ware and hotel ware; this firm is described in Chapter 9 as its main concern is the production of industrial ceramics.) Finally, the stoneware sector of the industry is represented by Denbyware Ltd. Many firms, just as successful and interesting and important in the industry as those included, have not been considered.* To include more would result in an undue amount of repetition, and also introducing too many names could lead to unnecessary confusion.

SPODE LTD.

A Royal Academy of Arts publication, *200 Years of Spode*, published in 1970 contained the following passage:

*The annual publication *Tableware Reference Book*, which is published by *Tableware International*, lists most of the manufacturers and their main products and agents.

The Spode tradition continues today as it has always done. Its essence lies in combining every effort to remain in the vanguard of new advance and development with an equal attention to maintaining the standards of the past and providing a continuity with it. This link with tradition is very real. The factory stands on its original site in the heart of Stoke-on-Trent; many of the original buildings have been preserved just as the Spodes knew them; many of the work people represent the third and fourth generations of families devoted to producing Spode. The finest collection of antique Spode in the world is located within the factory. In such surroundings it is not difficult to cherish the heritage of the past but here, too, is collected evidence of a constant concern with the present and future. A few yards from the old Spode bottle oven, preserved as a monument to Spode potters of other days, stands some of the most up-to-date gas-fired kilns in the world. Within the workshops of the Factory, the latest machinery and equipment is in use much of it invented or developed there, like the Murray Curvex printing machine which enables prints to be applied directly to the ware from copper engravings—the greatest single advance in the ceramic printing process for a century.

Josiah Spode II is credited with being the first manufacturer to make standard English bone china. He entered into partnership with William Copeland who operated the firm's London warehouse and in London they traded as Spode and Copeland. Between 1867 and 1970 the firm traded as W. T. Copeland and Sons Ltd. In 1966 the firm was acquired by Carborundum Ltd., an American company. Throughout the firm's long history, top priority had always been given to maintaining the high quality of its products. This continued to be the very essence of the company's policy and the reason for its continued high standing in world markets. Over the years the products and the factory in which they were made acquired a romantic image in the minds of customers, particularly in North America. The fact that Spode was still being produced in the original factory founded by Spode I was a valuable part of this image, but the legacy of old buildings scattered over a large site proved an increasing economic disadvantage requiring a large outlay of capital to remedy it. The firm had always kept faith with customers by producing ware at prices which were fair and reasonable. This also became increasingly difficult since so many of the traditional Spode designs could only be adapted to more economic forms of production by a very costly programme involving the most advanced techniques. The profit-earning capacity of the company thus became increasingly affected in recent years and it became evident that a takeover or amalgamation was inevitable in the climate of change which dominated the industry after 1950. The other members of the Fine China and Earthenware

Plate 5.1   Spode Ltd: An engraver at work

Manufacturers' Association, with the exception of Minton Ltd., were changing rapidly or being changed. Doulton, Royal Worcester and Wedgwood were changing rapidly and Royal Crown Derby had been swallowed-up by Allied English Potteries, leaving Copelands (later Spode) and Minton as firm favourites for takeovers by one of the 'Big Four'. The management team of W. T. Copeland and Sons Ltd. did not want to become part of a larger British firm as there would be the possibility of the identity of the firm disappearing in the long-run. Nevertheless, a takeover appeared to be inevitable as reconstruction and modernisation of the Spode Works was overdue and more money had to be spent on marketing activities, as even Spode wares could not be expected to sell themselves for ever. The offer by Carborundum was therefore a welcome one and it was accepted. The American Company quickly provided additional finance to modernise the factory (it was decided not to move out into the countryside) and a new management team was developed. Since 1966 reconstruction and modernisation has progressed rapidly, yet leaving the Spode attitude and philosophy virtually intact. Subsequently Carborundum acquired Hammersley & Co. (Longton) Ltd., a clear indication that a new pottery Group was being developed. A takeover bid for Aynsley China did not succeed. In 1970 Spode celebrated its 200th anniversary with appropriate dignity. It is reasonable to assume that eighteenth, nineteenth and twentieth century Spode designs will continue to be produced for hundreds of years to come.

MINTON LTD.

Minton Ltd., like Spode Ltd., is a firm renowned for the quality of its products. Our favourite story about Minton concerns our own University. The British Pottery Manufacturers' Federation were shocked to find the University in North Staffordshire entertaining distinguished guests off mere earthenware plates. They insisted on presenting the University with a fine china dinner service. The University stipulated that their dinner service should be white with a small version of the University coat of arms on the rim. The Federation provided five white plates, each made by a different manufacturer of fine bone china, and the University was asked to chose one of them. Seventeen visitors to and members of the University were asked to choose the best plate without knowing which manufacturer made which plate. With only one exception they all preferred the plate submitted by Minton.

Minton had avoided the introduction of machinery as much as possible for as long as possible, and the firm's experience in the post-war years was similar to that of Spode. Minton was founded by Thomas Minton who at one time had been employed as an engraver by Josiah Spode. Although Minton wares are held in the highest esteem, in 1960 the capital of the firm amounted to only £500 000 and profits were a mere £43 000. Presumably the directors of the firm had regarded growth as vulgar, something which might tarnish the perfection of Minton bodies, decorations and glazes. Finally, in 1968 Minton Ltd. was acquired by Doulton Ltd., and output has been increased. Extreme care has been taken to preserve Minton's reputation for dedicated production of top quality wares.

MASON'S IRONSTONE CHINA LTD.

In 1813 Charles James Mason of Lane Delph (now Middle Fenton) took out a patent for iron stone china. The patent specifies powdered slag of ironstone as one of its ingredients. Charles James Mason made all kinds of ware. Besides tableware, jugs, teapots and vases, he made large pieces such as fire-places, bed-posts, punch-bowls and foot-baths. The ware was strong and durable, and usually showily decorated. Mason may be commended for improvements made by him in enamelling and gilding earthenware. Like many of his contemporaries, Mason founded his decorations on Oriental models. Not for nothing had Miles Mason, his father, been a 'Chinaman' before he turned potter. Charles James Mason from 1813 worked in partnership with his brother, George Miles Mason, and after his retirement, with Samuel Baylis Faraday. He became bankrupt in 1848 and, at a sale which followed, the copperplates and moulds were bought by Francis Morley from whom they passed to the present owners, Mason's Ironstone China Ltd.*

Today, the company employs some 520 workers, including many part-time married women who are highly skilled decorators. The firm continues to produce Mason's richly enamelled patterns. The offices and factory are situated in the centre of Hanley; indeed, they have been on the same site since 1849 and are adjacent to Stoke-on-Trent's most modern cinema and bowling alley.

The Chairman, Mr. J. S. Goddard, places top priority on sustaining the high quality of the firm's products. The Managing Director, Mr. W. G. Evans, recognises the value of the prime objective for marketing and he ensures that the policy is effectively and profitably implemented.

Mason's dinnerware is sold almost all over the western world; the strong markets are in North America, Europe and the United Kingdom. It is particularly popular with sophisticated, well-to-do people who still regard opulent soup tureens, sauce boats and vegetable dishes as essential accompaniments of a main meal. Mason's shapes and decorations defy mechanisation to any great extent and their continuing success may be attributed among other things to a lack of low-priced substitutes.

The firm concentrates on traditional patterns and shapes, and although it is continually freshening its range of patterns with new reproductions and adaptations of old ones, it does not offer a series of new patterns every year. Its customers want traditional patterns that

---

*Two recent books are entirely devoted to Mason's Ironstone: R. G. Haggar, *The Masons of Lane Delph and the Origin of Mason's Patent Ironstone China*, Lund Humphries for G. L. Ashworth & Bros. (1952), and G. A. Godden, *The Illustrated Guide to Mason's Patent Ironstone China*, Barrie & Jenkins, London (1971).

only Mason's can supply. Of recent years there have been several periods when the world demand for its productions have far exceeded its capacity to supply, largely due to lack of the highly-skilled labour necessary. This would obviously be an intolerable situation for customers wanting contemporary ware, as fashions would change before deliveries were made. Because of the traditional nature of the ware Mason's find that their customers are prepared to wait for deliveries. Traditional ware and a stable demand appear to go together. In 1973 Mason's were acquired by Wedgwood Ltd.

STAFFORDSHIRE POTTERIES (HOLDINGS) LTD.

The company's main trading subsidiary is Staffordshire Potteries Ltd., which mass-produces and distributes earthenware tableware. In 1951 the company commenced manufacturing and trading at its present location, occupying at this present time a total area of 17 acres of a site which forms a part of the former Stoke City Airport at Meir on the southern edge of Stoke-on-Trent.

The business of six other manufacturing subsidiaries of Staffordshire

Plate 5.2   Staffordshire Potteries Ltd: Modern automatic cup-making machine (output exceeds half a million articles a week)

Potteries (Holdings) Ltd., which in the immediate post-war years occupied various sites in Tunstall, Hanley, Fenton and Longton, has gradually been transferred to the Meir site. Although some traces of these subsidiary companies can still be found at the production stage, all other functions, administration, marketing, etc. are now centralised and the Group trades entirely through one subsidiary, Staffordshire Potteries Ltd. There are also two selling and distribution subsidiary companies in Canada, Multiple Sales Ltd. and Staffordshire Potteries (Canada) Ltd., and an associate company in New York, English Dinner Wares Inc., which distributes and sells the company's products in the United States. Staffordshire Potteries and the Clough Group between them dominate the mass-produced inexpensive earthenware market.

The layout in the main production unit is rational. There is a direct flow of materials and ware from sliphouse to warehouse, and processes are linked by means of conveyor belts. One would think that this description would apply to most pottery factories; however, in domestic ware it is rarely so. Plants tend to accumulate bits and pieces which are fitted in here and there. A lot of labour time is in consequence taken up with fetching, stacking and carrying. In Staffordshire Potteries the clutter to be found in many pottery factories is absent,

Plate 5.3 Staffordshire Potteries Ltd: Automatic cup-making machines (materials and product are untouched by hand prior to handle fixing)

and all floors and benches are kept scrupulously clean. Once-fired ware constitutes a high proportion of the firm's output. By combining biscuit and glost firing in one operation, costs are reduced. Because it is particularly cheap, once-fired ware tends to be regarded as inferior. However, the salesmen of Staffordshire Potteries claim that by using the once-fired technique there is a greater fusion between the glaze and the body, and a more durable finished product is achieved. The easiest type of cup and mug to make by machine is the plain cylinder shape. Brightly coloured cylinders are apparently, from the point of view of Staffordshire Potteries, very much a fashion item which consumers want to buy. Keeping up with fashion is not only a matter of using recent shapes and colours from Carnaby Street, but the result of long term market planning, combined with the services of a competent design team.

Mr. E. C. Bowers, the Deputy Managing Director, like Mr. Clough, is an enthusiast about machines. The existing plant at Staffordshire Potteries reflects the company's ambition to create one of the best earthenware manufactories in the world. The rise in the Group's profit-rate in recent years reflects their success in developing and ironing out the teething troubles in newly installed automated equipment. Particular attention is paid to the needs of workers by eliminating, where possible, awkward heights, and avoiding the lifting of heavy weights. Pride in performance has not been lost in spite of potting skills having in large measure been transferred to machines. Pride in the well-designed and smooth-running series of production units has tended to compensate for skills lost; high wages generated by the mechanisation also help.

Mr. C. W. Bowers, the Managing Director, is an enthusiast about selling pottery just as his brother is an enthusiast about making it. The two brothers are regarded as having created around them a particularly strong management team by other managers in Stoke-on-Trent. The marketing side to the business has not been neglected, and lively design, low prices and good packaging have been used to advantage. In common with most other firms who manufacture earthenware, there is a pressure to 'move up' into better quality ware. This means achieving a superior finish and imposing stricter selection in the expectation of being able to command higher prices. Staffordshire Potteries have pioneered selling by means of self-service packs, and gradually they are having success with a difficult new technique. If they succeed in gearing supermarket selling to their production then they will do very well indeed. They have sold large quantities of cheap ware abroad, particularly in Sweden and Germany. It has been

Plate 5.4 Staffordshire Potteries Ltd: The latter section of an automatic plate-making machine (Looking from left to right can be seen: Automatic transfer of 'green' ware from mould to white dryer, automatic transfer of ware from white dryer to fettling machine, and automatic removal and stacking of finished ware)

argued that because all countries can make cheap earthenware, only British fine china and expensive earthenware would continue to find markets abroad. It has not worked out that way. Staffordshire Potteries have discovered growing markets for cheap mugs and other tableware in contemporary and traditional designs in high income countries. This has been achieved in the face of keen competition from Japanese and German firms in particular.

The existence of Staffordshire Potteries is a consequence of the foresight and initiative of Mr. C. G. Bowers. He recognised clearly that making and selling pottery for everyday use is a straightforward

Table 5.1 STAFFORDSHIRE POTTERIES (HOLDINGS) LTD. AND SUBSIDIARIES: FOUR YEAR RECORD

	1970	1969	1968	1967
Turnover	2 334 835	1 934 776	1 597 316	1 432 617
(Including export sales)	529 219	334 956	188 500	53 500
Trading surplus	249 411	196 915	140 977	97 996
Depreciation	53 807	49 161	43 746	40 565
Trading profit	195 604	147 754	97 231	57 431
Loan interest	20 000	20 000	20 000	18 789
Profit before taxation	175 604	127 754	77 231	38 642
Taxation	70 773	53 379	27 000	10 406
Available for distribution	104 831	74 375	50 231	35 280
Ordinary dividend paid	49 104	40 920	31 203	26 002
Profits retained	55 727	33 455	19 028	9 278
Current assets	876 071	756 838	548 108	435 640
Less current liabilities	733 736	647 367	501 264	399 095
Net current assets	142 335	109 471	46 844	36 545
Fixed assets	798 715	541 717	476 511	455 282
Investments	21 251	8 028	—	—
Less future taxation	72 000	53 000	27 000	12 500
Capital employed	890 301*	606 216	496 355	477 327
Profit before taxation (as % of capital employed)	22·0*	24·4	19·6	12·0
Average total labour force	1 170	1 145	1 074	1 002

*Due to revaluation of the land and premises and release of depreciation originally charged thereon, the capital employed for the year ended 30 June 1970 was increased by £178 358. Without this adjustment comparable percentage figures of profit before taxation to capital employed would have been (1970) 27·5%.

business operation and not a mysterious craft practised by a limited number of families steeped in the appropriate tradition. He persuaded firms to move from cramped sites and join with him to rationalise their production and marketing facilities before they experienced to the full the disadvantage of being small. Because of his strong convictions he took great risks. His sons subsequently consolidated the Group he had founded and developed it in new directions. Once again we find that a large pottery firm is flourishing as a result of the vision and energy of a founder who sought for changes that would result in sustained growth.

ALFRED CLOUGH LTD.

Alfred Clough Ltd. is a holding company. The Group includes two family firms which had been highly successful earthenware manufacturers between theWars: Barker Brothers Ltd. and W. H. Grindley & Co. Ltd. Old age finally caught up with the proprietors and their firms were purchased at what at the time appeared to be bargain prices. Barker Brothers was acquired in 1959 and W. H. Grindley in 1961. In 1955 Cartwright & Edwards Ltd. had been acquired; it was bankrupt at the time of the takeover and was a subsidiary purchased from a public holding company. It, too, had had a long history of being a highly successful earthenware producer. The Clough Group did not buy success by means of takeovers. The opposite was the case: the Group bought trouble in the expectation that new management and modernisation would bring success. As we shall see, the Group has had a long struggle against adversity, some of it of its own making; nevertheless, the Group is now firmly established. It may be seen that profits have fallen to low levels in recent years. There would appear to have been two main reasons for this—the Group's prices were kept too low for too long and parts of the Group made losses which offset profits made elsewhere. The factories which made losses were retained, however, in the expectation that, given time, new machines, equipment and buildings would obliterate the losses. The Group's original factory, the Royal Art Pottery, was purchased in 1938 and was finally sold in 1969. It, too, produced earthenware, tableware, teapots, vases, etc. The Group now consists of Cartwright & Edwards, where particularly cheap once-fired earthenware is mass-produced mainly for the home market, and Grindley and Barker Brothers, which between them produce medium-priced earthenware and hotel ware for home and export markets, exports accounting for some 60% of output. The firm's products are well-designed and manufactured to high standards, important ingredients of the Group's long-run competitive strength.

There is a marked contrast between the Clough Group and, say, Wedgwood or Doulton. The latter firms acquired only successful companies; they depended for growth on taking-over firms whose growth potentials had not been fully realised, usually because of management weaknesses which the acquiring companies expected to overcome. They believed they had surplus capacity in management which takeovers would eliminate. At the time of the takeovers the Clough Group did not always have adequate surplus capacity in management available and the management capacity obtained by the takeovers tended to be over-estimated. In contrast to Wedgwood or Doulton, the Clough Group purchased companies that either had failed or were about to fail. The reasons why the purchases were made were: (1) the replacement value of the assets acquired were regarded as some four times the purchase price; (2) workers were acquired at times when labour was particularly hard to get—without takeovers expansion would have been held back by a lack of labour; also some managers were acquired who were expected to be invaluable to the Group; (3) new market areas were opened up: the firms tended to have good order books; and (4) it was expected that new directors and new thinking would quickly transform losses into profits. The aim was profitable growth, and it was considered that growth would be achieved more quickly and permanently by means of takeovers than by means of internal expansion. The takeovers were in line with economic principles—choose the path of expansion which will produce the highest return. Of course, much uncertainty can be involved in takeovers and subsequent events may reveal the original expectations to have been misplaced. Luck is also involved, so a successful outcome could be a result of bad judgement and vice versa. As we shall see, luck at times deserted the Clough Group.

Another contrast between this Group and the others is that it (the Clough Group) was a family business (at present it is gradually changing to a more impersonal type of business organisation). Mr. Alfred J. Clough was the Chairman of both Grindley and Cartwright & Edwards, and Chairman and Managing Director of Barker Brothers until his retirement in 1972. His son, Mr. A. R. G. Clough, is the Managing Director of Grindley, Edwards and Barker Brothers, and other members of the family contribute to the business.

It is the energy and enthusiasm of Alfred J. Clough for making pottery which brought the Group into being and helped to sustain it.

In turn Mr. Clough regards his father as the true founder of the Group. He had started business as a pottery factor as long ago as 1905, and his business expanded considerably before World War I. In 1919 Mr. Alfred Clough joined his father's business straight from school. Subsequently Mr. Clough managed a number of factories for his father, ending up as production manager of the Royal Art Pottery in Longton in 1938.

Mr. Alfred Clough is a large man. When he was in charge he was restless and aggressive. He is liked and respected in North Staffordshire for his toughness and his humanity and his determination to grasp a larger share of the tableware market. He once explained that you could depend on machines, they never tire, whereas workers sometimes fail to turn up and even become ill! His passion was always to make more and better pots and to talk about manufacturing processes in a pub in Longton or in his Club in Stoke-on-Trent. It is reported that when he could tear himself away from his factories he would ride a horse to work-off surplus energy. Earthenware was the thing he really knew about and he understood and delighted in machinery. He made frequent trips to Germany, France and Italy to study semi-automated cup and flat ware machines. He would pick up used machines for a fraction of the cost of new ones and would work with his team of fitters to integrate them into his production units. His prime objective was to sustain a high level of production, thereby achieving a low level of unit cost. This in turn provided the sales staff with the advantage of offering particularly low prices. To obtain the best out of the machines the operatives are offered bonuses for achieving output targets and regular attendance. The consumer was offered familiar and competently made tableware at low prices and, because of economies of scale and relatively low expenditure on management services, profits were earned as a consequence of particularly low costs. The profits in turn are used to pay off the overdrafts inherited when companies were acquired and the overdrafts raised for capital expenditures on additional machines, equipment and buildings. So the success of the Group depended on adequate finance being generated to finance the modernisation of the four companies. This depended on costs being held down by means of mechanisation and this, in turn, depended on how successful the managers under Mr. Clough were in achieving high levels of output from the plant obtained by means of the takeovers combined with additional machines purchased subsequently.

His first takeover, that of Cartwright & Edwards, went very much according to plan for the first 10 years. Subsequent acquisitions, however, brought serious and difficult problems. Each new factory required development plans at the same time as the modernisation programmes for the original factories were maturing. The time needed to install and bring up to full production 'new' plant was sometimes underestimated. At times machines failed to fulfil their makers' promises, or deliveries of machines and spare parts were delayed. Most important, the Group did not recruit and retain an adequate number of really good managers. (To the extent that good managers usually generate more net revenue than they earn, it can be a mistake to economise by employing too few or inadequately qualified managers.) The Clough family were forced to spread their talents too thinly over the various factories, and delays and breakdowns seriously interrupted the flow of cash into the Group that was needed to sustain expansion on a wide front. It would seem that one firm too many was acquired. The Grindley takeover brought many problems which have taken years to overcome and which, in turn, seriously interfered with routine management in other parts of the Group. One may conclude that only firms with adequate management teams should normally be taken over; firms which do not meet this requirement should be taken over only at well spaced out intervals of time.

The 1960s were extremely busy years for Mr. Alfred Clough. However, he always had worked a six and a half day week and a ten hour day, so he was accustomed to it. His energy and enthusiasm and that of other members of his family has meant that the Group has survived. It now contains three modern and well-equipped factories and new management teams have been developed. The various units operate fairly independently, with specialist services supplied centrally to all the Companies—laboratory facilities, design, costing, and the design and development of machines.

It should be noted that the output of the Clough Group is restricted to earthenware and there is a big commitment to cheap earthenware in Cartwright & Edwards. The Group therefore operates in the most competitive sector of the industry, where profits are particularly difficult to sustain. A high degree of mechanisation is one way to do it. However, as has been pointed out many times, the mechanisation of clay using processes is not easy and takes a lot of time.

Figure 5.1 shows how the profit rates of the two major firms in cheap earthenware production varied in the 1960s. It appears from the figure that rarely did both firms do well in the same year. (The Clough Group also makes substantial quantities of middle-range and high quality earthenware, so the comparisons, strictly speaking, are not restricted to cheap earthenware only.) It is perhaps not surprising that

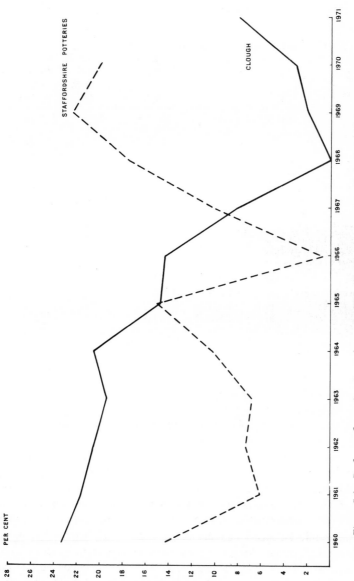

Figure 5.1 Profit rate fluctuations in earthenware: The Clough Group and Staffordshire Potteries 1960–1971

in 1972 it was reported that a major shareholder in Staffordshire Potteries had made a takeover bid for the Clough Group. In the event he did not succeed. In 1973 a merger was arranged between the Clough Group and Hostess Tableware.

Table 5.2   ALFRED CLOUGH LTD. RECORD 1955–1971

Year	Profit (£)	Employment	Turnover (£)	Capital expenditure (£)	Return on capital (%)
1955	40 000	300	250 000	5 000	28
1956					34
1957					38
1958	83 000				29
1959	114 000				30
1960	152 000	2 000			23
1961				106 000	22
1962					20
1963				120 000	20
1964	180 000			43 000	21
1965	138 000				15
1966	140 000	1 520	1 800 000	300 000	14
1967	106 000	1 260	1 703 000		8
1968	dr. 33 000	1 384	1 806 000		–
1969	27 000	1 404	1 857 000		2
1970	50 000	1 210	2 077 000		3
1971	144 000	1 210	2 370 000		8

J. & G. MEAKIN LTD.

In the early 1960s the weakness of top management in the tableware sector of the pottery industry was frequently discussed. The weakness was usually traced to a lack of good marketing—making products and then looking for customers rather than finding out what customers really wanted and then making it—and Doulton and Wedgwood would be mentioned as firms that were practising what the self-styled experts in business administration were preaching. Only a few other firms would be mentioned and invariably one of them was J. & G. Meakin Ltd. It was a long-established and respected family firm which, in the early 1950s, had been making losses. Here we concentrate on what happened after two men, J. W. E. Grundy and A. Derek Jones, 'knowing nothing about management and little about potting were appointed to the board as Chairman and Deputy Chairman in 1957'.

This account finishes with the merger with W. R. Midwinter Ltd. in 1968. Subsequently both companies were acquired by Wedgwood Ltd. (In a sense being taken-over by Wedgwood is a clear indication of success as they avoid buying failures even when they appear to be cheap.)

The fact that a new management team made a number of seemingly obvious and sensible changes that changed losses into profits and stagnation into growth suggests that in the past management had been ineffective and inefficient. Certain important aspects of management had been neglected; even so, they had operated by pioneering the use of machinery and paying close attention to overseas markets. The company had been built-up by the proprietors devoting long hours to manufacturing consistently good quality ware in a wide variety of shapes and sizes at prices customers could afford. The low cost of labour helped considerably. The following quotations are taken from *The Story of J. & G. Meakin 1851–1951* by Bernard Hollowood.

Long before the term was used in its present sense the brothers James and George Meakin built a factory that embodied all the principles of *rationalisation*, and the success of their venture stimulated similar advances throughout the industry . . . George Meakin set an example of missionary zeal which has been faithfully followed by the firm's directors ever since. His task, when he set out for America in the early 1850s, was to promote a flourishing trade between Hanley and America: and he succeeded. He soon discovered what designs would please the peoples of the Middle West, and it was largely from his prompting that a highly successful range of shapes and patterns was evolved. . . . Some years before World War II the firm embarked upon a massive scheme of reconstruction . . . in step with these developments were the reconstruction and electrification of the mill, the introduction of semi-automatic making-machines and an entirely new system of internal transport (by monorail, pipe-line, conveyor belt and self-propelled trucks), the erection of a large engineering department and the extension and modernisation of the research departments.

The quotations suggest that the firm had always done the right thing, the needs of customers in a growing market were studied and production was mechanised and engineering services and laboratories had been developed. However, in the main the machines in the new buildings had been used to speed-up non-machine processes. It had not been sufficiently realised that machines should be operated as near as possible continuously on long-runs of standardised patterns and that this in turn required a revolution in promotion and distribution. After 1957 new policies were adopted.

Mr. R. Fletcher was appointed to act as production manager. As well as knowing about production planning, he had been a successful salesman in a number of firms and industries. (It was like engaging Gary Sobers to revive a cricket team.) In addition, a number of managers already with the firm were good at their jobs and welcomed new ideas and new methods. Mr. Fletcher explains that, before he agreed to take the job, he demanded that a much larger biscuit warehouse space should be made available. His condition was accepted and it was then possible to remove ware waiting to be decorated and/or glazed from passages, working spaces, offices and the delivery warehouse. Later unsold stocks which had accumulated over the years were either dumped or sold to dealers by the hundredweight. Ware had accumulated because orders had not been achieved or had been cancelled, also because replacements might be asked for, and as more and more patterns had been introduced few had been abandoned.

The new team of managers met regularly in the firm's private dining room. A rigid rule was imposed: that lunches should be leisurely affairs and that 'shop' should be avoided. It must have been a terrible strain for potters to discuss the death of Queen Anne, Coronation Street or the Impressionists while orders were not being finalised or delivered or a kiln was misbehaving or piece rates were not being fixed. (The strain was not so great on the Chairman as, being a solicitor, he had plenty to talk about other than pots.) However, the benefits were substantial. Clear-cut decisions had to be made in a shorter time and more responsibility had to be delegated to middle managers than was normal in the industry. A habit of thinking strategically was developed which quickly provided the firm with a reputation for efficiency and flair. The policy that was hammered-out was based on a thorough understanding of the market and prices and production methods and costs.

Delivery was a problem. In the 1950s and for most of the 1960s most tableware firms had long order books for their successful patterns and delivery could be delayed for anything from a month to two years. Delays caused great concern to retailers; ware would be ordered when it was in fashion and delivered when it was out of fashion; no wonder retailers complained. Part of the trouble lay in the extremely wide product range offered by each firm; always some patterns were being delayed as others were being rushed through. As production was usually for orders received and not for stock the process tended to be self-perpetuating. In spite of the tendency for demand to exceed supply (a clear indication that prices were determined by costs and not by demand) firms tended to operate for short periods on short-time. The loss of a substantial overseas order might be the cause. Firms would produce for stock only to a limited extent because warehouse capacity was severely limited and there was uncertainty about future demand for particular patterns. Meakins recognised the benefits that would be obtained if delivery could be guaranteed within a few weeks when orders were placed, and the way to achieve this end was to restrict severely the product range. Between 1959 and 1966 Meakins cut the number of decorated patterns in production from 119 to 89. This does not appear to be a severe reduction; the significant thing was that the 89 patterns were based on only four basic shapes. Just as in the motor car industry, it was found that a wide variety of output can be achieved by using only a few mass-produced standardised components. One more important step was taken to simplify the product range: patterns were classified into categories so that sales and promotion could be related to a series of patterns and not to particular patterns. This introduced a degree of flexibility and aided forecasting as it was easier to guess what sales were likely to be for the group than for a particular pattern. Also promotions were geared to the category or group and this greatly extended the time they could be kept going. The 'Bull-in-the-China-Shop' range was particularly successful; consumers recognised, because of advertising in magazines, on television and in shop displays, a familiar and friendly bull and Meakin's sales benefited.

The success of the policy depended on well-chosen designs. Firms tended to offer 20 to 30 new patterns each year, usually at the Blackpool Fair, and it was left to retailers to indicate the popular ones by the size of the orders they placed. After a period the patterns with only few orders would be dropped. This system had two weaknesses: first of all it offered the consumer little say in the choice of new patterns and consumers were offered what the retailers believed they would like. One consequence was that designs tended to be geared to the rather conservative taste of retailers. Secondly, production schedules were interrupted and runs were shortened by the policy of running a large number of patterns to discover the survivors by a process of the survival of the fittest.

The success of the policy of a restricted range of patterns depended on, among other things, really good designs. They had to be distinctive and good in the sense that consumers liked them. Designs were commissioned from outside consultants and the firm also employed its own designers. The designers used bold colours and simple shapes with reflected fabric designs which consumers found vaguely familiar. Consumers on the mass market tend to like tableware designs which do not break entirely new ground; however they welcomed a change from roses and violets. The final choice of all designs chosen for long production runs was made by the directors.

J. & G. Meakin's policy of investigating what the consumer wanted and, on the basis of this information, adding designs to their pattern groups to replace those that failed to fulfil expectations, resulted in economies in production. In addition—and this was important —pressure was placed on retailers to stock Meakin patterns because they had been brought to the consumers' attention and they were likely to ask for them by name. From the point of view of retailers the big attraction was delivery made within three weeks of the receipt of orders. At last pottery was coming into line with almost all other consumer goods. Special pattern groups were developed for export markets, in particular for Canada and the United States, and the firm's export manager and other directors paid regular visits to their agents to see that retailers were provided with the full range of planned services.

The usual reasons given for delays in delivery were labour shortages, delays in the delivery of lithographic transfers, coffee pots or sauce boats not produced in sufficient quantities to complete sets, or kiln and machine breakdowns. Meakin took steps to minimise the impact of factors that could prevent delivery promises not being fulfilled. The personnel department was strengthened in order to minimise the disruptive effects of absenteeism and labour turnover. The firm's training programme was improved and working conditions and piece-rates were improved to attract and retain labour. Lithographic and other pattern transfers were ordered well in advance in sufficiently large quantities to prevent shortages arising, and new machines and kilns were installed and care was taken that they were properly maintained. There is no doubt that the firm's workforce responded favourably to the initiative displayed by the new managers.

Finally, prices were quoted which tended to be a little above those quoted for comparable tableware. This was possible because of the off-setting impact on demand of advertising and promotions, because retailers were being offered extra services in terms of delivery and attractive point-of-sale materials. Also, higher prices probably suggested to many consumers that Meakin's tableware was superior. It is just possible that sales were increased rather than reduced by the price increases. An attempt was made to avoid the image of cheapness which is projected by many manufacturers of tableware. Instead, an image of good modern tableware was projected which, because of its quality, had to be sold at higher prices. It was the pricing policy pursued by many manufacturers of consumer goods and which, at the time, was pursued by only a few pottery firms. The outcome of the new policy was lower unit costs and higher prices, and the result was an end of losses and a steep rise in profits; and profits were what the whole exercise was about.

When Meakin merged with Midwinter, they merged with a firm which had developed along similar lines. It was an earthenware manufacturer with interesting modern designs. J. & G. Meakin were not alone in cutting costs by severely restricting their product range. Mr. Kenneth Wild successfully pursued similar objectives when he was the Managing Director of Thomas C. Wild & Sons Ltd. (the firm is now part of Doulton/Allied English Potteries). In 1957 it was decided that 'instead of producing the hundreds of patterns which then came out of the factory, it would be better to rationalise to a few dozen well-thought-out designs with tested consumer reaction; in 1969 there are under 50 patterns'.*

J. & G. Meakin were pioneers in good management practice which was based on applying good common sense and keeping steadfastly to a policy. Other medium-sized firms in the industry either behaved in a similar fashion after 1965 or, as is likely, appreciated what Meakin had achieved and set about initiating similar policies.

T. G. GREEN LTD

T. G. Green Ltd. manufactures a wide range of earthenware with an emphasis on kitchen ware—bowls, jugs, mugs and basins. The factory is a collection of undistinguished buildings on a stretch of semi-derelict land near to Burton-on-Trent. This was an old establishment which was purchased by the present Managing Director, Mr. P. H. Freeman, in 1966. The purchase of the Company, which otherwise might have ceased to trade, was arranged by a finance company, Barro Brothers. New managers as well as new money were injected into the firm and close attention was paid to marketing in all its aspects, from improving designs to changing relations with wholesalers and retailers.

T. G. Green Ltd. has consistently earned high profits. In their considered opinion, they benefit significantly from being a medium-sized

*Kenneth Wild, *Pottery Gazette and Glass Trade Review* (April 1969).

firm (250–500 employees) and are convinced that they have the advantage in competing against Royal Doulton Tableware Ltd. or Wedgwood Ltd. By remaining of modest size they retain a high degree of flexibility and by avoiding the extra administrative costs which are inevitable in larger organisations they realise that they can compete vigorously with larger firms. Their Managing Director is convinced that takeovers and amalgamations have been harmful to the pottery industry; in his opinion they result in unduly high costs and prices and inferior service to retailers. However, not many small and medium-sized firms in the pottery industry have been as successful as T. G. Green Ltd. It is likely that there is an insufficient number of competent top managers to ensure that the industry could be dominated by small rather than by large firms. Inevitably some competent managers aim for growth and size and small firms become amalgamated into larger ones. It has been demonstrated by T. G. Green Ltd. that, by keeping its product range under strict control and relating prices closely to the requirements of various markets and sustaining its reputation for quality and service, costs can be held down and markets penetrated in spite of competition from larger rivals. The success of T. G. Green Ltd. may be explained by the firm's ability to resist temptation: it has not been tempted to expand by acquiring other firms and it has resisted being taken-over by another firm.

THOMAS POOLE & GLADSTONE CHINA LTD.

Thomas Poole (Longton) Ltd. and Gladstone China (Longton) Ltd. were family businesses, manufacturers of medium and best quality bone china dinner and tea sets. In 1948 the two firms amalgamated. Thomas Poole sold their china under the well-known trade name of Royal Stafford China and it has been retained by Thomas Poole and Gladstone China Ltd. The two companies combined constituted a modest-sized firm; the capital employed was £318 000 in 1954 and £313 000 in 1969, while employment in 1970 was 300 persons. In 1956, 1959 and 1968 no profits were made and after 1956 the highest rate of profit earned on capital was only 9%. One reason why profits were not satisfactory was the firm's persistence with an unduly wide range of patterns which prevented long production runs. Unit production costs were therefore high and prices could not be raised because there were fairly well-determined prices given by the market for conventional patterns. In 1961 the goodwill of Salisbury China Co. Ltd. was acquired; the effect of this acquisition was to increase the range of patterns even more. It was recognised that there is a limit to what can be done with bone china only and that it would be desirable to offer customers earthenware as well as china. In May 1970 the British Anchor Pottery Co., an earthenware manufacturer, was acquired from the Gailey Group for three million Poole & Gladstone shares at par, approximately £150 000. Although it was a takeover by Poole & Gladstone, it resulted in the Gailey Group being the major shareholder. A new Chairman was appointed, Mr. J. Nash, an accountant who heads a number of companies outside the pottery industry, including the Gailey Group of companies. Presumably a radical reconstruction of the management of Poole & Gladstone was planned before British Anchor was 'acquired'.

The new management team gave top priority to investigating the world market for bone china. Their conclusions were surprising. First of all it was decided to cease to give priority to the Canadian and United States markets. There was strong competition there from Japanese and German manufacturers and, more important, from large British manufacturers whose names were known to the buyers; American buyers were reluctant to place orders with an unknown firm. Although it would be possible to sell large quantities of ware in America, it would be difficult to earn any profit on the transactions. Europe offered Poole & Gladstone an alternative market. It was unusual for French and German customers to have heard of many British tableware manufacturers, so Poole & Gladstone felt that they would not be at a disadvantage in competing there with other firms that are well known to pottery buyers in the United States. By and large, in Europe, and especially in the European Economic Community countries, markets had been neglected by British pottery manufacturers. Plenty of scope was found for selling traditional British tableware patterns on the Continent. Distribution was rendered particularly simple by arranging for large importers to supply the whole of their market and refusing to sell from the factory direct to retailers. Europe also offered a limited scope for contemporary designs in earthenware, designs which are normally restricted to the home and Commonwealth markets, and more important, there seemed to be scope in Continental markets for contemporary designs in bone china—a range of products virtually neglected by competitors. It was also found that new markets could be found in Commonwealth countries for patterns developed for the Continental market. The decision was taken to concentrate on Europe. A gamble was involved: that if Britain should join the Common Market then, unlike most British pottery firms, Thomas Poole & Gladstone could be firmly established there. British membership would mean that as tariffs between Britain and member countries disappeared they would probably increase in the United States and appear in the Commonwealth. Europe not only offered scope for exports in the present, it was a market that was likely to expand considerably for British tableware in the future. Now that Britain is a member of the European Community the firm's pioneering efforts are likely to result in handsome dividends.

The next step of the marketing exercise was to develop new contemporary shapes and patterns for both earthenware and bone china. This was done for each main market segment—wedding presents, the middle-aged and elderly—and output and sales rapidly increased. Most of the patterns that had accumulated by the three companies over a period of 80 years were eliminated and longer runs substantially reduced costs. The new and traditional patterns are marketed under the new trade name Hostess Tableware. Arrangements were made for a manufacturer to supply vitreous enamel saucepans, etc. in colours to match the Hostess range of tableware. By emphasising fashion it is hoped that replacement demand will be stimulated.

In September 1971, an investment company was acquired with interests in domestic cutlery and table linen. This acquisition means that Poole & Gladstone can diversify its interests out of bone china and earthenware. By maintaining only financial links between producers of pottery, cutlery and table linen, the dangers of managers attempting to undertake too many technical and commercially unrelated tasks are avoided. It is unusual for a pottery firm to diversify its product range to the extent planned by Poole & Gladstone. However, there are precedents. Both Doulton and Wedgwood have acquired interests in glass, cutlery and jewellery. Pyrex markets a wide range of kitchen utensils along with its toughened glass tableware and ovenware and the Prestige range of kitchen utensils has been extended by the acquisition of an earthenware manufacturer.

The sale of the original Gladstone factory, with its picturesque bottle ovens intact, to a Trust that will convert it into a pottery museum is symbolic of the radical changes that occurred recently in a long-established firm that lost its momentum in the 1950s. Finally, in 1973, a merger was arranged between Poole & Gladstone (Hostess Tableware) and the Clough Group.

ORNAMENTAL WARE

Three firms which manufacture ornamental ware, Arthur Wood &

Son (Longport) Ltd., Portmerion Potteries Ltd., and Elijah Cotton Ltd., specialise in making and selling modestly priced wares. They are all small firms and, as is inevitable, each is dominated by policies dictated by one man (in the case of Portmerion, one woman). They may be regarded as being fairly representative of a sector of the industry where small firms persist. Beswick Ltd. was a larger and highly successful manufacturer of ornamental wares before it was acquired by Doulton, and the Royal Adderley Floral Co. is now part of the Allied English Group of companies. Both companies tend to supply ware that is more expensive than the ware supplied by the three companies considered here. In turn, Royal Crown Derby, Wedgwood, Doulton, Spode, Royal Worcester and others tend to supply even more expensive, and possibly, though not necessarily, better ornamental ware. The three firms share two things in common: they are all located in Stoke-on-Trent and they manufacture earthenware only. Many firms which are considerably smaller exist which manufacture, in the main, stoneware; they do not, however, cater for a mass market. In contrast, the profits of our three firms depend on their selling substantial amounts of every item they produce. They are in business for commercial reasons and not merely to combine the pleasures of making artistic pieces of pottery with earning an adequate or even inadequate income.

ARTHUR WOOD & SON (LONGPORT) LTD.

The Managing Director is Mr. A. F. Wood, who joined the family business in 1958. The firm had been founded by his grandfather, who had concentrated on the production of brown teapots. Subsequently his son (Mr. A. F. Wood's father) added a wide range of fancy earthenware to the teapots. The development of the firm in the 1960s may be regarded as a consequence of the attitude of the newcomer to the business: 'Either I make big changes and improvements or I get out.' This attitude, of course, is adopted by most businessmen. The family tradition and the desire of the workforce for continuity constituted additional pressures to stay in the industry and succeed in it. The desire to make money was not paramount in the decision to contribute to making more and better ashtrays, sweet dishes, vases and spice jars, as capital could have been transferred to alternative uses.

The firm's Directors recognised that with rising labour costs it would not be possible for a small firm to compete directly with larger ones in the production of conventional tableware; instead it was essential to specialise in ware that had unique characteristics and which provided scope for prices to cover costs and yield profits. A young designer was introduced to the firm to work systematically on a new product range and additional designs were commissioned·from various Colleges of Art. A policy of selling mainly to wholesalers was abandoned. Direct contacts were made with retailers who normally sell only small quantities of pottery and glass—gift shops, newsagents, variety stores, etc. Selling is conducted in a sector of the economy that does not normally impinge on the pottery industry. Competing products range from get-well cards and bracelets to cushion covers and pieces of brass. Products must be promoted in ways that would appear unconventional or even perverse to tableware manufacturers. A successful producer of fancy earthenware must consistently be aware of the luxuries the mass of the people, teenagers in particular, are most likely to purchase in addition to the essential luxuries of travel, entertainment, tobacco, drink, furniture and clothing. The success of the business has depended on the management keeping closely in touch with what is new and possibly profitable, as reflected in what is stocked by gift shops, particularly what is imported, and what is being featured and discussed in women's magazines. Continuously old products are being replaced by new ones and established products are being improved. The firm, up to the present, has not attempted systematic market research; instead, a cheap and possibly adequate substitute is found in the combined flair of the Managing Director and his designers and salesmen.

Two takeovers strengthened the firm considerably. Price and Kensington Potteries Ltd. were acquired in 1934 and their designs and patterns augmented those of Arthur Wood & Son Ltd. Carlton Ware Ltd. was acquired in 1967 and this enabled the firm to offer better quality and more expensive lines to a public that was tending to be willing to pay more for ornaments. Three separate sales forces have been retained on the principle that three sales forces are better than one. In the 1960s profit rates were fairly respectable for the pottery industry, ranging from 12·2% in 1966 to 18·4% in 1967 and 12·8% in 1969. The capital employed in 1969 was £360 000 and turn-over was £698 000. The main problem faced by the firm remains that of continuously improving a product range which will command profitable prices. Brown teapots tend to cost more to produce than consumers are willing to pay and their output has been severely curtailed. The following extract has a familiar ring: 'Production costs continue to spiral, particularly wages and materials, and this is bound to have an adverse effect on profit unless the company can recoup extra expenditure by advancing selling prices. Every possible endeavour will be made to do this but it is not easy in overseas markets where the company has to meet international competition.' It is taken from the 1970 Chairman's review.

ELIJAH COTTON LTD.

When Elijah Cotton died in 1895 his firm was known throughout the trade as 'Cotton's for Jugs'. He had built his business up from virtually nothing: his skill, hard work, judgement and a few borrowed sovereigns. His widow engaged a manager to keep the business going and eventually her two sons took over. Gradually the product range was extended to include vases, flower holders, dressing table accessories, ewers and basins and a limited range of coloured and banded tableware and kitchenware. Export markets were developed in Australia, New Zealand and Canada. Toward the end of the 1950s Elijah Cotton Ltd., in common with many other family businesses in the pottery industry, found it increasingly difficult to cope with rising labour costs; in particular, the firm's product range was not properly tuned to changes in consumer wants. Changes in the containers for beer and milk had severely contracted the demand for jugs.* In 1964 Mr. S. S. Hodson was invited to join the company as Managing Director and new products and methods of selling were rapidly introduced to remedy a deteriorating situation.

Mr. Hodson's management experience had been obtained in industries manufacturing consumer products, and latterly in the cutlery industry, an industry in which, like the pottery industry, small family businesses are important. The new Managing Director also had had experience organising courses on management in Commercial and Technical Colleges. His new post was therefore treated as a do-or-die exercise in marketing skills. A Marketing Manager, a Cost Accountant and a Production Controller were appointed (Mr. Hodson retaining responsibility for export activities) and the new management adopted a target to double the sales of the Company in terms of money over the five-year period 1966–1970. (Most books on management recommend 'Management by Objectives' as good business practice). At the end of 1970 the firm's target was exceeded by £18 000. One advantage enjoyed by the new management team was the trade name 'Lord Nelson' which had been approved by Lord Nelson in 1956.

* It was reported in the *Evening Sentinel* on 10 May 1957 that Mr. William Johnson 'placed a chum in his jolly mould and made what is estimated to be his 10 millionth jug. . . . His employer, Elijah Cotton Ltd., regard him affectionately, as one of those old-time loyal workers (now, they feel, becoming rare) who are a credit to their employers and an example to the younger generation'.

It had been discovered that pottery had first been made on the firm's premises in the year when the first Lord Nelson had been born.

Where possible direct representatives were engaged to replace agents in the United Kingdom, and existing wholesale distribution supplemented by a policy of searching out and opening first class retail establishments. The latter were given the full backing of 'Gondola' type store display stands and point-of-sale display material.

Market research was undertaken by Head Office and sales staff in the United Kingdom and by the firm's principal agents in foreign markets. The firm now manufactures six main categories of products: (1) jugs for the home, for catering establishments, hospitals and government contracts; (2) traditional and modern tea, coffee and dinnerware; (3) kitchenware which includes baking sets, salad and hors d'oeuvre sets, storage jars, etc.; (4) beer tankards; (5) fancies and boxed giftware, which include jewellery pendants and pomanders—a small jewellery manufacturing department is part of the factory; (6) sets which combine hardwoods and stainless steel with pottery such as cheese boards and toast racks. The new product range has been devised to minimise seasonal fluctuations in production for the home market. A special effort was made to provide prompt delivery and warehouse capacity was extended to this end and it is used only for wares that have good prospects of early sales. The managers believe their new product range provides them with a high degree of flexibility so that they can cater promptly for changes in taste and fashion.

In 1971 the firm had 300 employees and sales were in excess of £500 000. In addition to greatly increased turnover (in 1971 exports were up by 62%) the management are satisfied that the profit yield will prove as satisfactory as all other aspects of the business. The Company today differs from what it was in three main respects: (1) products are geared closely to market prospects: they are no longer produced without particular market segments in mind; (2) the company has moved much closer to consumers by supplementing wholesalers' activities and completely new outlets have been developed with chemists and variety stores; (3) delivery times have been shortened and prices are based partly on accurate costings which ensure that every line contributes to the profitability of the business.

PORTMERION POTTERIES LTD.

Small firms tend to prosper by catering for markets that are neglected or overlooked by large ones. By being flexible in their approach, small firms can not only supply specialised and neglected markets but also develop new markets by developing new wants. It is sometimes claimed, mistakenly, that the large pottery groups will eliminate the small firms in the domestic sector of the industry. This is a most unlikely possibility in an industry where catering for changes in fashion is important. Portmerion is a good example of how rapid growth and a secure position can be achieved by a small firm directed by a management team with ability and clear objectives. The Chairman of Portmerion Potteries is somewhat unusual in being a woman (Susie Cooper makes two). Miss Susan Williams-Ellis (Mrs. J. Cooper-Willis) and her husband took over two very small firms, Kirkhams and Grays, in Stoke-on-Trent in 1961 and called their new firm Portmerion (there were family connections with Wales). The products of the firm have a lot in common with craft pottery: it is earthenware and not stoneware, yet browns and greens are more common than the whites so beloved by most British earthenware firms. The distinctive ware of Portmerion is well-represented in kitchen and art-and-craft shops as well as in the outlets that normally sell pottery. Portmerion, however, is not a craft pottery; it is highly mechanised, with a workforce of some 170 persons, and some 50% of output is exported. The old factory has been modernised throughout and new kilns and machines have been installed. Production has been backed by advertising and public relations and close attention has been paid to good packing and prompt delivery.

It is tempting to account for the success of the new firm by the Portmerion fresh and attractive designs (where else could one obtain a Florrie Ford plate?); however, the attention paid to work study, production control and the application of advanced accounting techniques (in the context of the pottery industry) should not be overlooked. The Chairman, who is a designer (assisted by her economist husband), has recruited a team which has made profits on the basis of supplying consumers with novel and attractive objects at moderate prices. The heart of the matter lies in the firm's well-designed and competently managed production unit, and its consistently market-orientated approach. The firm has also provided the design, marketing and sales services for Dartington Glass Ltd., which has been built-up with similar policies.

DENBYWARE LTD.

This group has two operating subsidiaries in the United Kingdom:

Joseph Bourne & Son Ltd. and Langley Pottery Ltd.; both companies manufacture stoneware. Millard Norman Co. is a subsidiary company in the United States; it distributes Denby and Langley wares and also imports related products. The group also owns 40% of the issued capital of International Ceramics Ltd. which produces ceramic cores which are used in the precision casting industry, e.g. turbine blades for the aircraft industry.

Plate 5.5   Denbyware Ltd: Potter at the wheel

Tableware manufacturers occasionally display attractive patterns which have won design awards and remark: 'They look nice, but they were flops, the public refused to buy them. It is essential if profits are to be made to offer insipid and even ugly and bad designs if that is what customers will buy.' At the factory in the village of Denby, near Derby, there is obviously no conflict. Shapes, textures and designs are what they are because they are kept strictly in line with the traditional aesthetic standards of craft pottery. The ware is strongly influenced by contemporary fashions in fine art. The market has reacted favourably to the firm's refusal to compromise with a lack of appreciation of good design on the part of consumers. (Consumers who believe that well-designed plates and bowls should be thin, normally white, and decorated with pure unadulterated colours, are precluded from admiring good quality stoneware.) It could be argued, however, that only relatively few families purchase stoneware; in comparison with mass-produced earthenware it is expensive, so Denby can pursue a policy

Plate 5.6 Denbyware Ltd: Making a casserole

that is denied to firms that cater more for the mass market. Even so, the sales by Denbyware in 1970 were £2·6 million which included exports of £600 000. Sales had more than doubled since 1966.

Stoneware is only a small fraction of the total tableware output. Sales, however, have grown much faster than those of bone china and earthenware since 1948 (*Figure 5.2*). For years Denbyware Ltd. had the growth very much to itself. Recently they have been joined by Wedgwood and Doulton and other firms too offering stoneware oven-to-tableware.

A visit to Denby creates an impression of a competently managed

factory. In relative isolation a high degree of goodwill has been
established with the labour force of some 700 persons. The ware is
made from clay taken from the hill on which the factory is situated.
The clay is unique and all Denby pieces are hand-painted and there is

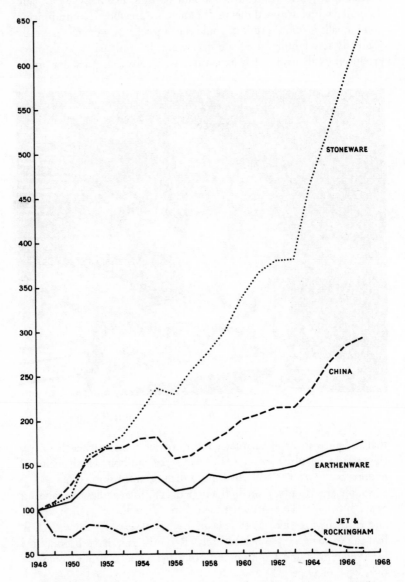

Figure 5.2   Comparative growths of sales of types of tableware (1948 = 1968)

even a cluster of potters' wheels kept in constant operation. There is plenty of space and light in the factory and cup-making and plate-making machines of advanced design have been installed. The ware is fired in modern electric kilns.

Proper attention is paid to marketing in all its aspects. Output is for stock and new lines are introduced and old ones retired systematically. The product range is wide with over 15 000 different products. There are 25 Denby shops-within-shops. Sales are promoted mainly by advertising in magazines. Because the firm's tableware and ornamental ware is distinctive, prices are sufficiently high to provide good profit margins; a firm control of costs also contributes to this end. Profits have enabled the firm to grow and invest in adequate production and marketing facilities.

6

COSTS

... about the year 1887 printing on glost by a process called chromo-lithography began to be introduced firstly by Mons. Pierre Rataud, and has made great strides; whereas formerly poly-chrome had to be filled in by hand in the tedious method of pencilling-in of each flower. Now, by a dab, the whole spray, say of a delicate and highly artistic design, can be adjusted momentarily. Add to this the recent introduction of liquid gold, which does not require any burnishing, as was formerly the case, to brighten it—these improvements and others have tended to lower the expense of the manu-facture to a marked degree without deterioration to the article itself, nay, one may assert an amazing progress.

W. Scarratt, *Old Times in the Potteries* (1906). Republished by S. R. Publishers Limited, 1969.

The cost structure of an industry is of interest for several reasons. Firstly, in a broad way it can show the relative contributions (measured in terms of rewards received) made by labour and capital employed within the industry, and by the various suppliers of raw materials and other inputs.

Secondly, the relationship between costs and prices is an important aspect of business behaviour and decision making, which shows the way in which technological factors on the one hand react with the market on the other. In addition, because technology, the market for the product, and the markets in which factors of production are pur-chased are in a state of continuous change, the cost structure is likely to change in response to these. Furthermore, there is considerable variation between products of most of the factors mentioned, but since most of the available data relates to the whole industry or large sub-sections of it, such variations will be dealt with in a qualitative rather than quantitative manner. Yet another aspect worthy of attention is a problem common to all multi-product businesses, that of allocating costs to particular products, and this too has repercussions for pricing policy.

THE STRUCTURE OF COSTS

Pottery has traditionally been a labour-intensive industry. This was partly because of its craft nature, but also a consequence of the factor-markets in which it operated. On the one hand, labour was plentiful and cheap because of the limited alternative sources of employment, which were coal mining for men and virtually non-existent for women. On the capital side, because of the intensity of competition which prevailed, profits were low, and with the family firms of which the industry was largely composed, profits were the main source of investment capital. Thus it was both unnecessary in terms of factor prices and difficult in terms of resources for the industry to become more capital intensive until changes in the labour market and in technology brought about an acceleration of the pace of investment. The growth of new industries in North Staffordshire and the transition of the country as a whole to a full-employment economy were responsible for the former, while the latter involved the use of much larger units of capital, particularly in kilns. This extra investment was financed in various ways; some from higher profits earned during the post-war period, some from the public flotation of companies pre-viously privately owned, and a substantial portion by means of capital made available when pottery firms were taken over by firms in other industries.

An idea of the degree of labour intensity in pottery compared with other industries can be obtained from *Table 6.1* which shows the ratio of wages and salaries to value added for a number of industries in 1963. It can be seen that the pottery industry as a whole has as high a ratio as any other industry and higher than most, and also that the industry's average conceals considerable variation between sectors, so that for the domestic ware sector the ratio is considerably higher. It might be argued that such ratios are open to misinterpretation since what is measured is the total payments made to labour and capital,

Table 6.1 UNITED KINGDOM MANUFACTURING INDUSTRY; RATIO OF WAGES AND SALARIES TO VALUE ADDED, DISTRIBUTION BY INDUSTRIES 1963

Pottery	0·69	Chemicals and chemical products	0·35
Pottery (domestic & ornamental		Petroleum & coal products	0·40
ware)	0·75	Non-metallic mineral products	0·55
Food	0·44	Basic metals	0·57
Beverages	0·30	Metal products†	0·55
Tobacco	0·25	Manufacture of machinery ‡	0·58
Textiles	0·57	Electrical apparatus, machinery	
Clothing, footwear & made-up		appliances and supplies	0·56
textiles	0·61	Transport equipment	0·61
Wood & cork products	0·62	Other manufacturing	0·53
Furniture & fixtures	0·65		
Paper & paper products	0·50		
Leather & leather products*	0·56		
Rubber products	0·54		
Average all industries	0·53		

Sources: P. J. Loftus, *Lloyds Bank Review* (April 1969) and *Report on the Census of Production*, H.M.S.O., London, Section 103 (Pottery) (1963). (Elsewhere in this chapter the latter will be referred to as the *Census*.)
*Except wearing apparel.
†Except machinery and transport equipment.
‡Except electrical machinery.

which is a function of the quantity employed and the rate of return; but as long as there is reasonably free entry to an industry, the rate of return on capital is unlikely to remain substantially above the general level in the long run, and similarly if wage rates are persistently high in relation to the cost of capital one would expect a certain amount of substitution of capital for labour.

The ratios in *Table 6.2* reinforce what has been written elsewhere

Table 6.2 COMPOSITION OF NET OUTPUT OF POTTERY INDUSTRY, 1958 AND 1963

Sector of industry	Ratio of wages and salaries to value added	
	(1958)†	(1963)
Whole industry	0·68	0·69 (0·65)*
Domestic & ornamental ware	0·71	0·75 (0·71)*
Tiles	0·65	0·61 (0·57)*
Sanitary earthenware	0·63	0·63 (0·59)*
Electrical ware	0·61	0·66 (0·62)*

Source: *Census* H.M.S.O., London (1963).
*Excluding Employers National Insurance and pension scheme contributions.
†The ratios for 1958 do not include Employers National Insurance and private pension scheme contributions. These are included as a component of wages and salaries for 1963 but to facilitate comparison with 1958, 1963 ratios calculated without including employers contributions are given in brackets.

with respect to changing technology. For tiles and sanitary ware the ratio fell between 1958 and 1963, indicating an increase in capital intensity in these sectors, whereas for domestic ware the impact of increased mechanisation was later and capital intensity did not change. The apparent increase in labour intensity for the electrical ware sector is almost certainly a result of the volatile level of capital utilisation achieved in that sector, which, according to these ratios, was the most capital-intensive part of the industry in 1958. It is to be expected that the ratio for the domestic sector has now begun to fall, and when the detailed results of the 1968 census are available this will be able to be examined further.

Table 6.3 enables us to examine the general structure of costs when raw materials, fuel and other inputs are included.* The changes in the share of total value of output by various types of cost reflect, among other things, the technological and structural changes which were taking place during the 1950s and 1960s. We have shown above that the industry has become more capital intensive, and the wage and salary component of gross output fell from 44·1% in 1954 to 42·3% in 1963 if employers' national insurance and private pension scheme contributions are excluded, as they were in the census for 1954.

As has been mentioned already, there are almost certainly variations in the cost breakdown shown in *Table 6.3* between sectors and differences in the growth rate of sectors, leading to changes in their shares of total output, would of course cause the average costs to change. For example, the slight increase in the share of value of gross output taken by purchases of all kinds could be accounted for in terms of the more rapid growth in the tiles sector, since it can be seen from *Table 6.4* that 43% of gross output was spent on such purchases in the tiles sector in 1963, compared with 33·7% for the industry as a whole. This would also account for the diminished importance of some of the items not normally used in tile manufacture, such as bone and bone ash, and, to a lesser extent (i.e. some tiles are decorated) lithos.

The relative importance of fuel and electricity costs declined, but probably of more interest is the massive change in the composition from almost one-half coal in 1954 to under one-sixth in 1963.

It should be noted that *Table 6.3* relates only to purchases, and that there are some costs not included and therefore it is not possible to deduce from this data the gross profit as a share of output. In 1963,

* Unfortunately the detailed Census data relates only to the whole industry and obviously conceals possible variations between sectors and firms. The BCMF conducts an annual survey of costs on behalf of its members, but the results are available only to member firms, and then only in summary form.

Table 6.3 PURCHASES BY LARGER FIRMS IN THE INDUSTRY IN 1954* AND 1963, UNITED KINGDOM

Purchases	1954 Value (£)	1954 Cost of purchases (%)	1954 Value of gross output (%)	1963 Value (£)	1963 Cost of purchases (%)	1963 Value of gross output (%)
Materials for processing:	11 027	65·3	19·6	14 141	55·1	18·5
Flint	906	5·3	1·6	808	3·1	1·1
China clay	636	3·7	1·1	814	3·2	1·1
Ball clay	541	3·2	1·0	807	3·1	1·1
Other clays and marls	389	2·3	0·7	257	1·0	0·3
China stone and Cornish stone ⎫	} 537	} 3·1	} 1·0	452	1·8	0·6
Feldspar ⎭				339	1·3	0·4
Bone and bone ash	537	3·1	1·0	427	1·7	0·6
Quartz, whiting and other quarry products ⎫	} †			182	0·7	0·2
Heavy chemicals ⎬				163	0·6	0·2
Frits (lead) ⎭				290	1·1	0·4
Prepared bodies	237	1·4	0·4	371	1·4	0·5
Colours and materials for colours	850	5·0	1·5	844	3·3	1·1
Glaze and materials for glaze not specified above	694	10·0	3·0	1 475	5·7	1·9
Lithos	665	3·9	1·2	943	3·7	1·2
Kiln furniture ⎫	} †			728	2·8	1·0
Refractory materials not specified above ⎬				166	0·6	0·2
Ware purchased for decorating ⎭				324	1·3	0·4
Lubricating oils and greases				32	0·1	—

Replacement parts for machinery and vehicles, and consumable tools bought as replacement	1 192	7·0	2·1	1 032	4·0	1·4
All other materials for processing	2 842	16·8	5·1	3 687	14·4	4·8
Packaging materials	1 256	7·4	2·2	2 016	7·9	2·6
Fuel and electricity:	4 585	27·2	8·2	5 878	22·9	7·7
Coke and manufacturers fuel	258	1·5	0·5	23	0·1	—
Fuel for road vehicles	87	0·5	0·2	71	0·3	0·1
Other liquid fuels	192	1·1	0·3	731	2·8	1·0
Gas	1 168	6·9	2·1	2 408	9·4	3·2
Electricity	818	4·8	1·5	1 713	6·7	2·2
Total cost of materials and fuel ‡	16 866	100·0	30·1	22 033	85·8	28·9
Goods purchased for merchanting	—			3 397	13·2	4·5
Canteen purchases	—			232	0·9	0·3
Total cost of purchases	16 866	100·0	30·1	25 662	100·0	33·6
Wages and salaries	24 706		44·1	33 971		42·3
Gross output	55 979		100·0	76 336		100·0

*The 1958 Census did not collect detailed data on purchases.

†Not separately recorded.

‡Sub-totals may not sum to totals because of rounding.

Table 6.4 BROAD BREAKDOWN OF COSTS BY SECTOR* OF FIRMS EMPLOYING 25 OR MORE PERSONS (1963)

Costs	Whole industry		Domestic and Ornamental ware		Sanitary Earthenware		Tiles		Earthenware	
	Value (£'000)	Gross output (%)	Value (£'000)	(%)	Value (£'000)	(%)	Value (£'000)	(%)	Value (£'000)	(%)
Gross output	76 336		35 781		10 588		17 164		10 239	
Purchases:										
Materials for processing and packaging, and fuel	22 033	28·9	9 768	27·3	2 815	26·6	5 997	34·9	2 577	25·2
Goods for merchanting and canteen purchases	3 629	4·8	801	2·2	663	6·3	1 389	8·1	535	5·2
Payments to other organisations for work done or materials given out and for transport	1 668	2·2	396	1·1	406	3·8	561	3·3	219	2·1
Wages and salaries (including employers' contributions)	33 971	44·5	18 128	52·3	4 194	39·6	5 697	33·2	4 567	44·6

Source: *Census*, H.M.S.O., London, Table 2 (1963)

*The sectors listed do not account for the whole output of the industry.

78·1% of the value of gross output is accounted for by the costs included in *Table 6.3*. It is possible from other tables in the 1963 Census to identify the main costs not already included. These are summarised in *Table 6.5* and together account for 5·2% of gross output.

Table 6.5 ADDITIONAL COSTS NOT INCLUDED IN TABLE 6.3

Costs	(£'000)
Transport	
Payments to other organisations for transport	1 618
Costs (other than fuel and depreciation) of operating road goods vehicles	59
Payments for other services	
Repairs and maintenance to buildings	220
Repairs and maintenance to plant and machinery	717
Rates	852
Hire of plant and machinery	119
Postage, telephones, telegrams and cables	382
Total	3 967

Source: *Census*, H.M.S.O., London, Tables 11 and 12 (1963)

IMPORT CONTENT

The pottery industry makes a significant contribution to exports, not only in terms of the high proportion of its output which is exported, but also because the import content of that output is very small. Data enabling us to estimate the import content of pottery are provided in the *Input–Output Tables*.* For the purposes of these tables Pottery and Glass are aggregated together, so that we can only obtain an approximate result for Pottery. For a gross output valued at £228·4 millions in 1963 the combined industries purchased directly £3·4 millions of imported goods and £3·1 millions of imported services, but this does not include indirect imports, i.e., those purchased by firms supplying the Pottery and Glass Industry with raw materials and other inputs. If indirect as well as direct imports are taken into account, then the import content of Pottery and Glass was £89 per £1 000 of output available to final buyers, or £20·3 millions for the combined industries in 1963. If we assume that the import content of pottery is the same as that of glass and that within the pottery industry the import content of exports is the same as for the output of the in-

Input–Output Tables for the United Kingdom, H.M.S.O., London (1963).

dustry as a whole, then we can obtain an estimate that for 1963 the pottery industry's exports of £21·4 millions contained only £1·9 millions of imports, thus making a net contribution of £19·5 millions to the country's balance of payments.

MANAGEMENT AND COSTS

A central problem for many multi-product firms, and there are no pottery manufacturers who are not multi-product firms, is the allocation of shared costs. How important, then, are shared costs for pottery manufacture? Most direct process labour is paid using some type of piece-work scheme, so that this category of wage costs can be fairly easily allocated to particular products. In addition, there are some departments which are fairly specific in their function, and the costs of these must obviously be shared only among those items to the production of which they contribute. For example, nobody would attempt to justify allocating the cost of decorating processes to ware which is not decorated. In this respect the pottery industry has much in common with the chemical industry, and to a certain extent agriculture, in that some departments produce output which can be either intermediate and used as the input to a further process, or final and ready for sale. Once a piece of ware emerges from the glost kiln it is a potential final product, and whether or not it is processed further will depend on a number of factors such as, on the market side, the relative attractiveness of undecorated and decorated ware and, on the production side, the capacity for decorating in relation to making capacity, which in itself may vary depending on the particular product. In addition the degree of elaboration of the decoration is capable of considerable variation. This is not to say, of course, that in a pottery one would expect to find managers scanning the contents of the glost warehouse and deciding what will be decorated and what sold immediately; production planning is now carried out well in advance and the matching of the capacities of different parts of the factory can often be achieved by making such decisions only in relation to sub-standard ware.

In almost any organisation there are some costs which even the most ingenious cost accountant would find difficult to classify and allocate, and in pottery these general overheads are comparable to those found elsewhere. General management and administrative costs, plus those relating to research, design and selling, will almost certainly be recovered by the addition of a mark-up with respect to total allocated costs. In pottery manufacture there are in addition a number of functions, such as body preparation, firing, and the sorting and selecting of ware, which take place at several stages and which contribute to the making of almost every item, but for which one might initially consider there to be no obvious basis for allocation. (The result of the selection operations is that at a number of points in the production process, 'seconds' are being distinguished, and as well as having to allocate some or all of the cost of producing them on to the first quality ware which remains, the management team has also to decide on the most profitable way of disposing of them. This problem will be discussed in more detail later in this chapter.)

The clay preparation and firing costs mentioned above can in fact be allocated on a fairly reliable basis; in the former case the weight of clay used for each piece of ware is usually known, and in the latter the volume of kiln space occupied by different types of ware is known. However, since a proportion at least of the costs of operating the clay preparation department and the kilns are effectively fixed, any rule for allocating them is dependent for its effectiveness on accurately estimating the level of capacity utilisation which is likely to be achieved and at which costs are intended to be fully recovered. Thus the cost of preparing one ton of clay for the making department can be estimated on the basis of past experience and, of course, it will be known how much of this represents wages paid to slip house workers; and so if there were a wage increase it would be possible to calculate its effect on the cost of clay production, and similarly for a change in the cost of clay.

If one accepts that the slip house produces prepared body and the kilns provide a particular volume of firing capacity, then these departments are almost unique in that they can be said to have a homogenous product. They are, together with the biscuit warehouse, also unique in that they contribute to the cost of making almost every article produced by the firm. It is important for the remaining departments that as high a proportion as possible of their operating costs be recovered specifically on those items which have been processed within the department, rather than by recovering them on the output of the factory as a whole. A common way of achieving this is to calculate, historically, for each department, the ratio of total running costs to piece-work payments, and when costing an article to recover costs in each department at a piece-work rate (obtained by comparison with similar articles and in consultation with work study officers) plus an additional percentage of the piece-work rate to cover all other departmental costs. This allowance will almost certainly include payments to indirect labour, the cost of consumable tools, etc. and power costs, leaving only such costs as rates and rent, general administration and selling expenses, to be recovered as a 'general overhead'. An example of cost-plus pricing is given in Appendix 4.

Handling and warehousing costs can be added either to the making or firing process which precedes them or, more approximately, recovered as a rate per dozen for all items processed. The cost of sub-standard ware is recovered as a percentage addition at each stage corresponding to the percentage of ware rejected.

The next problem is the methods of recovery of the overheads not already allocated. There are firms which calculate the ratio of overheads to allocated costs for the factory as a whole, and add the resulting percentage to the allocated cost of each item. Alternatively the overhead figure could be related to direct labour costs, and it has been suggested to us (by a management accountant working with a consultant rather than within one firm) that a case can be made for relating the total overhead costs to the full capacity throughput (measured as a cubic capacity) of the biscuit kiln and recovering on this basis. This is justified as an aid to determining the optimum product mix, since in many firms the biscuit kiln represents the main production bottleneck; therefore products which occupy a relatively large amount of scarce kiln space will be shown up as less profitable than those which are less intensive in this respect, with the result that, the market permitting, the product will be changed. This argument is, of course, only important if the kiln is working to full capacity. It has already been mentioned that one of the disadvantages of the tunnel kiln is its relative inflexibility and the fact that its operating costs are virtually fixed, as long as it is being used at all. The increased use of multiple intermittent kilns instead of one tunnel kiln, particularly for glost and enamel firing, means that a much higher proportion of firing costs become variable and impose less severe penalties in terms of increased cost for operating below full capacity.

In the section on individual firms (Chapters 4 and 5) we have referred to the cost reductions achieved by firms as a result of rationalising product ranges and reducing the number of individual lines produced. Clearly, the type of exercise outlined above for allocating costs is carried out on the assumption that machines and production lines are in continuous operation throughout working hours. The greater the variety in the product line the longer will be the time spent adjusting machines, etc. and in addition the manager's task of planning production will be more complicated. Warehouses for work in progress will also be more difficult to operate. The number of machine changeovers is basically a function of two variables: the number of products for which a particular machine or production line is used, and the size of the run or batch when any item is made. Formerly the first of these was high and has been reduced by firms carrying out fairly drastic pruning operations on their product range. The size of each run tended to be small because production was to order rather than for stock, as indeed it had to be, since to produce such a variety of products for stock would have been out of the question. Once variety was reduced, however, the scheduling of long production runs of lines destined for the final warehouse rather than immediate despatch became a possibility, and firms such as Royal Albert, J. & G. Meakin and Midwinter reaped considerable benefits from such rationalisation exercises. Of course, the significance of rationalisation of the product range is not confined to production matters; in many ways break-throughs of the type achieved were more market orientated than production orientated, or at least demonstrated the extent to which the marketing function has become an integral part of the well-managed company's planning process. Another cost-significant aspect of the size of the product range is that variations can be achieved by varying either or both the shape and the pattern, and it is frequent changes in the former which are more important as contributors to increased costs. Thus we find that a number of types of decoration are commonly available on each of the relatively small number of shapes produced in any one factory. This concentration on a single basic shape with variations available in the superficial and peripheral aspects has its parallel in the sanitary ware sector, where one of the contributory factors to the cost of water closets was the wide range of angles and sizes of outlet, which was cast as an integral part of the piece. Rather than attempt to gain acceptability within the building industry for a reduced range of this type of variation it has now become common for firms to produce a standard item to which a variety of outlet adaptors (often not made of clay) can be fitted. Thus the cost of making and maintaining multiple sets of complex moulds has been eliminated and production planning considerably simplified.

One further aspect of the importance of allocating costs correctly to different products lies in the setting of relative prices for the different items in a set. Different types of costs may change over time at different rates, and the impact of advances in technology will be felt at different times for different processes. It is therefore important for firms to be willing to increase the price of some items more than others if this is required. If all items were sold in sets of standard composition this would be less imperative, and this perhaps accounts for the fact that during the 1960s the standard way of quoting export prices for earthenware was in terms of a base price plus a percentage, the base price originating from the earthenware exporter's trade association. But within the basic scales were contained relative prices for different items and types of decoration, and we were told that these relative prices were based on the relative costs as they were in about 1910. Such a system of quoting prices was justified on the grounds of the complexity of most firms' product ranges; and the replacement of this by firms' individual pricing procedures is yet another benefit achieved as a result of the product rationalisation exercise discussed above, and of the more widespread use of cost accounting methods enabling the relative costs of items to be known much more reliably than previously.

THE 'SECONDS' PROBLEM

The general problem of sub-standard output is, of course, widespread in industry; but in the pottery industry it is different from most industries in at least two respects. Firstly, it is very difficult and frequently impossible to convert a sub-standard item into a best quality one, whereas, for example, in light engineering an article which has been incorrectly assembled can be dismantled and reassembled, although the cost of doing so by hand may not always be justifiable. Secondly, there is a market for sub-standard ware. In fact, the manufacturer has three alternative ways of disposing of any ware which is not so badly distorted or damaged as to be completely useless. He can sell the undecorated ware to a wholesaler specialising in the distribution of seconds, or he can sell to one of the firms which buy seconds and decorate them for resale. Or thirdly, depending on the situation within the firm, it may be worthwhile to retain sub-standard ware and decorate it, possibly using up stocks of lithos for designs no longer in the product range. These decorated seconds can then be sold through a wholesaler or perhaps distributed to china shops, etc. in readiness for sales. In addition, it must be remembered that some faults may not occur until the article is almost finished, so that there will also be seconds which are fully decorated with the normal pattern. These also find their way into the January sales, and some factories contain a 'seconds shop' where ware is sold to employees, visitors and those members of the general public who are aware of their existence. It is sometimes said that no one in Stoke-on-Trent buys best quality ware, and although this is an exaggeration there is a good deal of truth in it.

The extent to which firms retain their own seconds for further processing will depend on the opportunity cost of doing so. A firm will not use up scarce enamel kiln-capacity firing decorated seconds if they are to occupy space which could be used in firing best quality ware, but at slack times more seconds will be retained and decorated. It is also possible that in some firms the standard of selection employed in the sorting warehouses will be varied in response to the state of the market. In addition, when selection for decoration is taking place, it may be possible to use more categories than simply 'accept' and 'reject'. For example, if one shape is being decorated with two types of on-glaze pattern, one type of decoration may be heavy enough to cover up minor faults in the glaze and pieces will be selected accordingly.

A number of interesting exercises have been carried out by firms to determine whether or not it is worthwhile to achieve reductions in the seconds rate. This is a complex problem, involving estimates both of the cost of reducing wastage and the extent to which sales of seconds are detrimental to the market for best quality ware.

7

THE HOME MARKET

DISTRIBUTION

Over the past 30 years there have been major changes in the way in which pottery is sold to the public. Whereas up to 1939 the bulk of pottery sales were made by glass and china shops, now only relatively few specialist shops remain in business. Glass and china shops have been replaced by china and glass departments in department stores and by variety stores such as Woolworths and Boots. Glass and china shops may now account for only 10%–20% of total retail sales of pottery.

In 1961 no fewer than 40 138 retail outlets were recorded which sold pottery and many more outlets sold some pottery but not enough to be recorded as sellers of pottery in the Census. Many of the small outlets were hardware stores, newsagents, market stalls, tobacconists, chemists, etc. who obtained their supplies from pottery wholesalers rather than direct from manufacturers. Pottery wholesalers supply retailers who sell only relatively small quantities of pottery, and retailers who operate in rural areas. Whereas before 1939 manufacturers sold most of their output to wholesalers who, in turn, sold to retailers, now manufacturers sell the greatest part of their output to retailers. The change in part reflects the higher degree of concentration in retailing as compared to pre-war; retail outlets are now controlled by a relatively small number of large retail organisations.

Many pottery wholesalers specialise in supplying the catering trade and they combine the sale of tableware and ornamental ware with the sale of bacon slicers, hot plates and chip pans. Some wholesalers act as brokers and undertake the whole of the selling side on behalf of small manufacturers. Some even arrange for biscuit ware to be decorated, so they supply their own exclusive patterns, and others specialise in buying seconds and export rejects and old stock for ultimate sale by market traders and by general retailers interested in offering bargains to the shopping public. Just as manufacturers sell direct to retailers, they also sell direct to hotels, hospitals, British Rail and air and shipping companies. In addition, contracts are placed direct with manufacturers of jugs, ashtrays, etc. by food and drink manufacturers.

MARGINS

Pottery manufacturers receive only some two thirds of consumers' expenditure on pottery. At least one third represents the costs of distribution. The mark-up on trade prices of pottery in 1968 varied between 40% and $66\frac{2}{3}$%.* The mark-ups are rather high for durable consumer goods. However, they probably are justified. The stock-turn on pottery is low, breakages inevitably occur and they adversely affect the sale of sets. Display is difficult and requires a lot of space and, to provide adequate service, shop assistants should know about body compositions, glazes and types of decoration. Replacements are a bother which do not affect the sales of saucepans, carpets or furniture to anything like the same extent.

DEPARTMENT STORES AND SUPERMARKETS

Department stores and chains of department stores are operated so that the rate of return on capital is the same for all departments. It is a rule that may be modified only in special circumstances. Glass and china departments are usually located away from the main entrances, in basements or on or near the top floor, where floor space is less costly that in more favoured locations. In this way china and glass departments can normally pay their way. Clothing, millinery, haber-

*National Board for Prices and Incomes, Report No. 55, Cmnd. 3737, Distribution Margins in Relation to Manufacturers' Recommended Prices.

dashery, jewellery, etc. are normally allocated the better locations as they can afford considerably higher rents and remain profitable. Location decisions inevitably affect the bonuses, etc. that shop assistants can expect to earn, so recruitment of competent sales assistants tends to be more difficult for china and glass departments than for (say) clothing departments.*

More and more expenditure by consumers is being channelled through supermarkets at the expense of the specialist shops and department stores. Without self-service it is extremely difficult to hold down retailing costs, and it is because specialist shops tend to be labour intensive that they have lost so much ground to the multiples in recent years. Unfortunately, from the point of view of pottery manufacturers, an insignificant amount of pottery is sold in supermarkets. From the point of view of supermarket operators, pottery takes up an undue amount of space relative to the value of sales and because of the numerous shapes, sizes, types and decorations of pottery, stock control can be unduly complicated. If the technical problems of selling pottery could be overcome, then impulse buying could contribute towards increasing the demand for pottery. Supermarkets would obviously be ideal outlets for pottery, as food and drink and pottery are complementary products. By means of consumer packs some pottery manufacturers (Staffordshire Potteries is one)† are achieving some success with sales through supermarkets. This could be the beginning of an important new development for tableware manufacturers.

STORES-WITHIN-STORES

In Chapter 4 it was noted that, among others, Wedgwood, Doulton and Royal Worcester operate stores-within-stores. This has been an important development, as it provides manufacturers with direct contacts with the public through their own trained staff who normally

---

*Some evidence that shop assistants in china and glass departments of departmental stores are less than perfect was obtained in 1969 by asking housewives their opinions of shop assistants. Of the householders in the sample 59% considered shop assistants selling china and glass in all retail outlets to be knowledgeable and 61% considered them to be interested. However, 78% of the housewives thought that assistants were in general helpful, and the majority of the sample considered that assistants in glass and china shops and on market stalls were both more knowledgeable and more interested than those in department and variety stores.

† See page 92.

operate the stores. It helps to ensure that shop assistants are knowledge-able about the ware they offer for sale. This is particularly important when sales are composed mainly of fine china and glass and oven-to-tableware. The stores-within-stores tended to cater almost exclusively for the well-to-do, and customers not dressed for shopping were inclined to prefer to do their shopping in a more relaxed and less refined atmosphere. In effect possibly only 70% of shoppers ever ventured into the stores and an even smaller percentage ever purchased anything. The character of the stores changed considerably after Johnson Brothers, J. & G. Meakin and Midwinter were acquired. Thereafter Wedgwood offered a complete range of expensive and inexpensive wares and the character of the stores changed so that rich customers were retained at the same time as not-so-rich customers were attracted.

The merger of Allied English and Doulton should result in a con-siderable increase in the price and product range of the Doulton stores. It is to be expected that the Lawley chain of shops, operated by Allied English, will improve and possibly expand as a result of the merger. A few glass and china shops have been established recently in main streets, shopping precincts and arcades by manufacturers; this is a worthwhile new development as it contributes towards halting the decline of the specialist shop. Part of the cost of running the shops could rightly be regarded by manufacturers as promotion expenditures. Excellent glass and china shops, and many of them, are obviously vital to ensure that sales of pottery in the U.K. increase in the future.

EXPENDITURE ON TABLEWARE

The outstanding feature of the expenditure on pottery in the home market has been its low level and the persistence of that low level. In 1970 the sales on the home market, at factory prices, were £32 million, made up of stoneware (£2 million), earthenware (£21 million), and china and porcelain (£9 million). The pottery is pur-chased directly at retail prices and indirectly at wholesale prices. Direct purchases are made mainly by consumers in shops and from market stalls, and indirect purchases are made on their behalf by catering establishments. The latter include cafés, hotels, canteens and refectories in factories, offices, schools and hospitals.

Firms manufacture hotel ware, which is usually more robust and less ornate than domestic ware; it is designed to stand-up to heavy concentrations of detergents in dish-washers, to stack compactly and not to chip or crack readily. Much of it is earthenware fired to higher temperatures than usual to produce vitreous ware. Whereas only small quantities of vitreous ware are purchased by households, fairly large quantities of ordinary tableware are purchased regularly by catering establishments.

We do not know how the market is divided between households and catering establishments; if we assume that one-third of manufacturers' sales are to catering establishments, and we allow for wholesalers' mark-ups of 15%, then expenditure by domestic caterers on pottery in 1970 was £12 million. (Actual expenditure includes a figure for imports.) If we add 60% to the remainder to translate factory prices into retail prices we obtain an estimate of consumers' expenditure on home-produced pottery in 1970 of £35 million. Total expenditure, direct and indirect, was therefore £47 million, a figure which, it must be emphasised, is subject to a wide and unknown margin of error.

It was shown in Chapter 5 (*Figure 5.2*) that between 1948 and 1968 the sales of bone china increased substantially more than the sales of earthenware. *Figure 5.2* also shows a rapid increase in sales of stoneware and a fall in the sales of jet and rockingham (black and brown glazed earthenware). It must be remembered, however, that stoneware only accounted for a small fraction of total sales in 1968 in spite of its rapid growth over the 20 year period from only a few thousand pounds in value in 1948.

The estimates of expenditure have been confined to home-produced pottery; however, households and catering establishments do not

Table 7.1 EXPENDITURE OF HOUSEHOLDS ON TABLEWARE AND ORNAMENTAL WARE BY MATERIALS IN 1967 (%)

Material	Total purchases	Own use	Gifts
China	27	24	35
Earthenware/stoneware	43	47	29
Plastics	5	7	2
Toughened glass	21	18	30
Others	4	4	4
Totals	100	100	100

Source: P. W. Gay and R. L. Smyth, *The Household Demand for Tableware in Great Britain in 1968*, Department of Economics, University of Keele (1971).

restrict their requirements of tableware and ornamental ware to pottery; they also purchase items manufactured from plastics, glass and toughened glass (Pyrex, etc.), and stainless steel. Even waxed paper, in the form of beakers and plates, is eating into the pottery market. *Table 7.1* shows how expenditure by households was distributed between materials in 1967. The table distinguishes between purchases for own use and purchases for gifts.

Table 7.1 also shows that plastics have not made substantial inroads into the market, whereas a substantial share of the market is held by the manufacturers of toughened glass. It is extremely expensive, if not impossible, to produce tableware in plastics which will not, over a period, stain or scratch. Manufacturers of plastic tableware have a secure market for picnic and camping ware, and they may succeed eventually in penetrating further into the more conventional tableware market. Their presence is a constant threat to china and earthenware producers and they indirectly contribute towards keeping pottery prices and profit margins low. In contrast, the manufacturers of toughened glass hold a substantial share of the market. The toughened glass manufacturers have been particularly successful in catering for oven ware needs and for oven-to-tableware. Recently there has been a fight-back by pottery manufacturers who have introduced mass-produced oven-to-tableware. From the table it can be seen that manufacturers of toughened glass, mainly Pyrex, have captured a large share of the gift market. The importance of competition between manufacturers of pottery, plastics and toughened glass, actual and potential, was emphasised in Chapter 3.

Expenditure on pottery in the home market increased in the 1960s mainly because prices increased. As it was a period of inflation the pottery price increases merely enabled manufacturers and distributors to take into account the general fall in the value of money. In real terms domestic pottery sales may have increased by about 1% per year in the 1960s. *Table 7.2* shows the gradual rise in the average weekly expenditure by all households between 1957 and 1969 and the accompanying rise in the wholesale prices of pottery. The volume index was at the same level in 1967 as it was in 1959, and this strongly suggests that the pottery industry is having to contend with a stagnant home market. In 1969 the figures show a sudden jump, in volume terms, from 111 to 160. Unfortunately, this figure most probably does not reflect what actually happened; it is likely to be the result of a large sampling error: 5·9 pence is an average figure with a standard error of 0·8, which means that the actual figure could be as low as

Table 7.2 FAMILY EXPENDITURE ON POTTERY AND GLASS, ETC. 1957–1969, AND
THE WHOLESALE PRICE INDEX OF DOMESTIC POTTERY

Year	(1) Weekly expenditure by all households (new pence)	(2) Standard error of (1)	(3) Index No. of expenditure	(4) Wholesale price index of domestic pottery	(5) Col. (3) ÷ Col. (4) Volume index of expenditure
1957	2·3		100	100	100
1959	2·7		117	105	111
1961	2·9		126	116	109
1965	3·3	0·4	143	135	106
1966	3·5		152	143	106
1967	3·8		165	148	111
1969	5·9	0·8	257	161	160

Sources: Col. 1, Family Expenditure Survey (Special Returns provided by the Department of Employment); Col. 4, China and Earthenware Wholesale Prices, *Annual Abstract of Statistics*.

4·3 new pence and as high as 7·5 new pence. The lower figure is in line with expenditures for previous years.

Family Expenditure Surveys are undertaken annually by the Department of Employment and Productivity and they include data relating to 'china, glass, cutlery, hardware, ironmongery, etc.' (item 66). The Department provided data relating to the chinaware, glassware, pottery, etc. element only of item 66, and *Table 7.2* was based on the special returns.* *Figure 7.1* and *Figure 7.2* are also based on Family Expenditure Survey data.

Figure 7.1 shows how household expenditure on pottery and glass increased in 1967 and in 1969 as income increased. The positive relationship between income and expenditure is significant; it does not occur by chance merely because of sampling errors. Similar relationships were found when selected earlier years were examined. Clearly income is fundamental in the context of expenditure. If the income of a household is unduly low then attractive tableware, and most other consumer goods, too, cannot stimulate expenditure.

Figure 7.2 also shows expenditure on pottery and glass increasing as income increased in 1967; in addition, it shows variations in ex-

---

*The estimates for the years 1957–1967 are presented in Appendix A of *The British Pottery Industry, 1935–1968* by D. J. Machin and R. L. Smyth (Department of Economics, University of Keele, 1969).

139

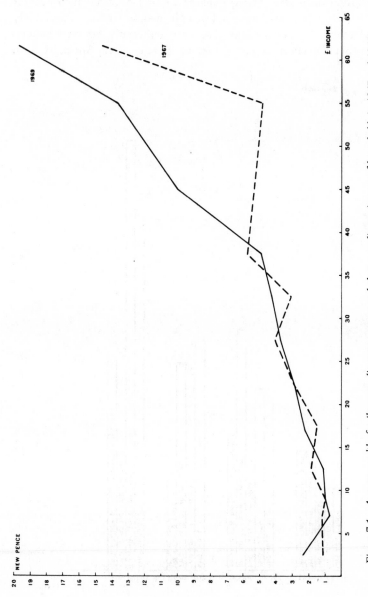

Figure 7.1   Average weekly family expenditure on pottery and glass according to income of household, in 1967 and 1969

penditure at various levels of income as family size varies. The contrast between' low income and high income families is interesting. Expenditure on pottery and glass increased as family size increased for low income families, but not for high income families. Presumably preferences for pottery and glass vary sufficiently between families to offset the effect of family size on purchases. In Appendix 5 an

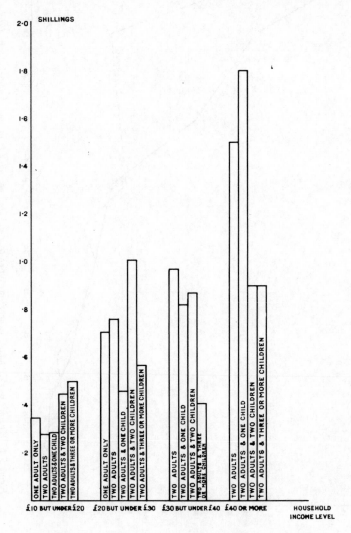

Figure 7.2 *Expenditure on pottery and glass by family size and levels of household income*

econometric study of expenditure on tableware demonstrates that family size is an important factor.

Figure 7.3 shows the relationship between total consumer expenditure and consumer expenditure on pottery since 1920. It shows that after 1955 both sets of expenditure rose. (The level of expenditure in the early 1950s may be explained by supplies being diverted from the home to the export market. Also, the post-war need to renew carpets, furniture, etc. probably held back expenditure on pottery until after

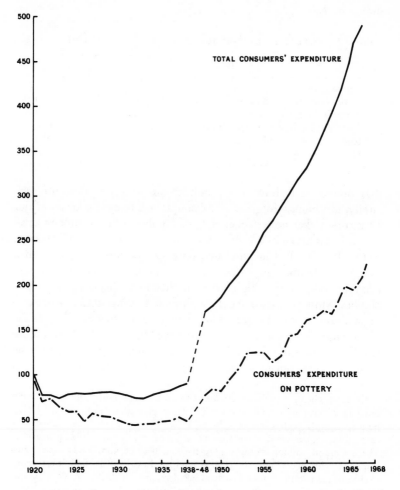

Figure 7.3 Total consumers' expenditure and expenditure on pottery: the United Kingdom 1920–1968

1955.)* The figures on which *Figure 7.3* is based suggest that out of every £1000 increase in consumers' expenditure, £2·1 is spent on pottery. Given that income and expenditure are likely to increase in the future, it is reasonable to assume that expenditure on pottery will increase too.

The expenditure on pottery in Britain is absurdly low; in 1967 it was four new pence per family per week, or approximately £2 per family per annum. Only part of the expenditure goes to the manufacturers, for the figure includes distribution costs, etc. In 1967 the weekly expenditure per family on various domestic items was as is shown in *Table 7.3*

Table 7.3   WEEKLY FAMILY EXPENDITURE ON DOMESTIC ITEMS IN 1967

Item	New pence
Pottery and glass	4·0
Cutlery, hardware and ironmongery	17·0
Furniture, including repairs	28·0
Soft furnishings and household textiles	17·0
Radio, television and musical instruments, including repairs	20·0
Floor coverings	23·0

A distinctive feature of expenditure on pottery is that, for each family, it is highly irregular. For example, we found by means of our 1968 survey that no less than 50% of families spent no money at all on earthenware and 60% of families spent no money at all on bone china. (If a family had purchased only an egg cup or a saucer then they were classified as having purchased pottery.) Each year the expenditure on pottery is highly concentrated in the hands of only a few families. Most families make do with their existing stocks of pottery. Expenditure is closely geared to family circumstances. Tea, coffee and dinner sets are a feature of most weddings, so that establishing families increase the demand for pottery. Later on, children may be expected to stimulate demand. Demand is subsequently likely to

*In 1958 when consumers were asked to rank certain household items in the order in which they would be most likely to replace them in their own home, pottery was placed last but one. In 1968, however, pottery came second to a carpet and was preferred to curtains, a canteen of cutlery and a mixing machine. The change in ranking strongly suggests that, because households were better equipped with everything in 1968 as compared with 1958, they were more likely to purchase more pottery after 1968 than after 1958. (See P. W. Gay and R. L. Smyth, *The Household Demand for Tableware in Great Britain in 1968*, Department of Economics, University of Keele, Ch. 8, 1971).

remain fairly dormant until the possibility arises of purchasing new pottery after the children have left the household. Demand tends to be held down by the determination of families to make their pottery last for at least one lifetime. *Table 7.4* summarises the age composition

Table 7.4 THE AGE DISTRIBUTION OF TABLEWARE (%)*

Maximum age	China	Earth-enware	Stone-ware	Plastic	Glass	Iron/steel	Total
3 years	24·7	42·9	45·9	66·9	48·7	55·4	39·7
5 years	37·0	58·0	65·7	84·1	64·7	70·7	54·0
10 years	57·2	76·5	79·3	95·9	85·0	90·6	72·9

*Each entry shows the percentage of the total stock of tableware made from the 'column' material which is within the maximum age stated for that row. Age was measured as length of time since purchase, not length of time since manufacture.

of the ownership of earthenware, bone china, toughened glass, plastic, iron and steel, and stone. No less than 63% of the bone china in households in 1967 was at least five years old. (This percentage refers to all bone china. It would be much higher if only fine bone china had been included.)

In 1967 the 800 families in the Keele survey purchased 14 982 pieces of pottery, etc. and owned 82 984. Ownership therefore represents approximately five and a half times annual purchases. *Table 7.5* shows

Table 7.5 CONDITION OF TABLEWARE ETC. BY MATERIAL

Condition	China	Earth-enware	Stone	Plastic	Toughened glass	Total
Very good	67·8	36·9	53·8	29·3	51·1	47·8
Good	28·7	47·8	36·8	57·9	43·9	41·7
Fair	3·4	1·9	0·4	1·3	0·5	2·2
Poor	—	12·9	9·0	10·6	4·2	7·9
Very poor	0·1	0·5	—	0·9	0·3	0·4
Totals	100·0	100·0	100·0	100·0	100·0	100·0

the condition of the tableware and ornaments owned, by material. Of these, 10·5% of the pottery was in 'fair', 'poor' or 'very poor' condition. If all fair to very poor ware were replaced entirely in one year, sales would in consequence increase by 58%. This suggests that the poor condition of pottery in households tends to be exaggerated and that sales are held down mainly by consumers being content with the patterns and shapes they own rather than by their using or merely keeping cracked and chipped crockery and ornaments.

One reason why expenditure on pottery is so low is that a wide variety of exceedingly cheap and serviceable tea, coffee and dinner oddments are supplied to the market by pottery manufacturers. Cheap pottery tends to drive out dearer pottery. The cheap pottery may be mass-produced once-fired earthenware or mass-produced bone china or seconds. Good earthenware and china is selected, which means that plenty of seconds are always available. Of course, much of the very cheap ware is purchased by poor families and by catering establishments who might otherwise purchase plastic tableware. Also, there is some stimulus to demand when cheap pottery is purchased. Consumers may be tempted to replace cheap ware by better quality and more expensive ware. It is claimed that many beer drinkers eventually become brandy drinkers and that sales of cheap small 'cigars' eventually result in extra sales of Corona Coronas. A similar argument applies to the sale of seconds and very cheap ware: they spoil the market only if consumers' preferences are fixed and incapable of change. It is not unlikely that consumers who purchase cheap pottery are eventually stimulated to purchase something a bit better, and the purchase of plastic beakers could lead to expenditure on bone china.

The ownership and purchasing of pottery tend to rise and fall together. This is a consequence of stocks being the outcome of a number of years of purchasing less breakages, and purchasing increasing as income increases. Families frequently purchase very little pottery, not because they own a lot but because pottery simply does not interest them. In contrast, there are families who regard pottery as providing more than just a bare household service and who own large stocks and are keen to add to them at frequent intervals. Also large stocks tend to stimulate replacement demand. When we asked a random sample of households in 1968 if they had 'too much', 'about right' or 'too little' pottery their answers were approximately one third to each category. Many low income families reported that they had too little pottery. However, there was not much they could do about it. Families which had stated that they had enough or even too much, presumably, would be inclined to make only infrequent and modest purchases. This brings us back to the beginning: the large purchases tend to be made by, or on behalf of, newly married couples. It would seem that a substantial stimulus to the demand for pottery on the home market will be achieved only when consumers decide to throw away the accumulation of years of purchases, and buy something different. Such action is unlikely to occur spontaneously. It will require a major marketing operation by manufacturers to change the attitudes to pottery of a large number of established families. It is the exceedingly low levels of expenditure which suggest that a sustained promotion programme could bring success.

Data relating to ownership and relationships between ownership and purchases are presented and analysed in Appendix 5.

We have attempted to explain the extent to which income generates expenditure and why the low level of demand for pottery persists. In particular we have noted the limitations of existing retail outlets and the difficulties involved in developing new ones. Greater activity on the part of wholesalers would not make much difference; because of the growth of direct trading with retailers, their role in distribution has contracted considerably in recent years. We will conclude this chapter by examining two more of the factors which contribute to the level of retail sales: advertising and other forms of sales promotion and prices.

PROMOTION

In 1968 housewives were asked 'have you been influenced at all to purchase tableware, ornamental ware or kitchenware by:

1. Advertising on television, in newspapers or magazines?
2. Displays in shops?
3. Sales or price reductions in shops?
4. Friends or the advice of friends?'

The answers given by a representative sample of housewives, in percentages, were as follows: Advertising 13%, Displays 40%, Sales 35% and Friends 20%. It is reasonable to suppose that if similar questions had been posed about washing machines, saucepans or carpets, advertising would have gained a much higher percentage. The low 13% for advertising strongly suggests that housewives in 1968 were only marginally aware of advertising by pottery firms. Presumably there has been some slight improvement in the position since that time. It would be a long period of time before a substantial increase in advertising expenditure resulted in increased sales and profits; nevertheless, without more advertising on television and in newspapers and magazines, it is difficult to visualise any substantial increases in demand. Manufacturers in other industries tend to promote their wares more actively than do pottery manufacturers and sales of pottery are in consequence adversely affected. Expenditure on advertising could be regarded as an investment. In the pottery industry investment has been concentrated on new machines, buildings and kilns at the expense of advertising. This could easily result in capacity increasing faster than demand. If profits had been higher, then perhaps more money would have been found for spending on advertising. It is understandable, if not entirely rational, that firms should economise on advertising, given that the benefits are difficult to isolate and measure.

Selling by many tableware manufacturers tends to be directed mainly at retailers and wholesalers rather than at final consumers (the force of this statement, fortunately, has been weakened somewhat in recent years by the growth of large groups). Sales representatives make regular calls on customers; in addition, showrooms for trade customers are a normal feature of each group of factories. A showroom may also be operated in London or a London wholesaler may act on behalf of the manufacturers. There is plenty of attractive advertising by pottery firms to be found each month in the industry's trade journal, *Tableware International*. However, it is intended to influence retailers and wholesalers and not the final consumer. Trade advertising helps to decide which firms obtain orders, but it probably does not contribute significantly to expanding the market for pottery. Similarly with trade fairs: they are directed entirely at retailers and wholesalers and final customers are excluded. The pottery manufacturers participate actively each year at the Blackpool Gift Fair. Perhaps the Fair, and other similar ones, contribute something to expanding the market for pottery; manufacturers at the Fair compete directly for orders with manufacturers of glass, jewellery and fancy goods of all kinds.

By means of a voluntary levy on manufacturers' sales the British Pottery Promotions Service Ltd. helped to stimulate sales by providing courses for shop assistants and talking about pottery in women's institutes, schools, etc. Only a limited amount of work could be performed on a budget of only some £8000. In 1970 the service was abandoned. This does not signify that pottery manufacturers are not aware of advertising and other forms of sales promotion, but rather that the larger groups prefer to undertake their own selling campaigns.

PRICING

The pricing of pottery is discussed at length in Appendix 4. It includes a number of statements by manufacturers on pricing policy.

A feature of the pricing of pottery in the post-war years has been the long waiting times for popular patterns. This is a clear indication that prices tend to be related to costs rather than to what the market would bear. Cost-plus or full-cost pricing is favoured because firms produce many lines and patterns and an attempt is made to ensure that each pattern bears an appropriate proportion of overhead costs and makes a proper contribution towards profits. Firms attempt to keep prices steady over fairly long periods so as to avoid changing too frequently the price lists they issue with their catalogues. In practice, however, the costs and profit margins need to be adjusted to ensure that the prices are kept in line with competing wares supplied by rival firms at home and abroad. In Chapter 5 it was noted how J. & G. Meakin decided to produce for stock and offer prompt delivery to customers. By backing their sales with advertising campaigns and other forms of sales promotion they were in a position to command higher and more profitable prices. There has been a tendency for firms to view their pricing as one element in a marketing exercise and to allow prices to reflect market forces rather than costs. This we consider to be a marked improvement on the use of crude cost-plus pricing formulae. Because of the strength of competition, manufacturers find that new patterns and designs must be introduced regularly to permit adequate profit margins to be earned. Some of the fine china manufacturers, in particular, have been successful in demanding profitable prices in return for supplying ware of a high quality which is distinctive.

8

EXPORTS

Figure 8.1 shows the value of exports from the United Kingdom of glazed tiles, sanitary ware, electrical ware and domestic ware each year from 1949 to 1971. The sustained growth of total pottery exports since 1952 is impressive. It clearly demonstrates that British manufacturers have a firm grip on export markets. It may be seen from *Figure 8.1* that domestic ware has consistently, year after year, accounted for by far the greater part of the exports of the British pottery industry. In 1963 the domestic sector accounted for 47% of gross output and 69% of the exports of the pottery industry. In the world market there is more scope for decorated tableware and ornamental ware than there is for sanitary ware or tiles or electrical ware. Countries tend to be more self-sufficient in industrial ceramics than they are in domestic pottery.

Figure 8.1 also shows that in the 1960s the value of exports of glazed tiles considerably exceeded the value of exports of sanitary ware from the United Kingdom. This was a considerable achievement and it was based largely on the development of light do-it-yourself tiles by the manufacturers which was backed by aggressive marketing. Although sanitary ware exports were less than tile exports each year after 1961, they were nevertheless sustained at some £3 million a year and increased after devaluation in 1967. This, too, was a good export performance in the face of increased protection in established markets; and new markets, particularly in the Middle East and Africa, were established. Exports of electrical ware were low in the 1950s because of the pressure of home demand; after home demand slackened it was difficult to penetrate into overseas markets. *Figure 8.1* shows, however, that exports increased after 1968.

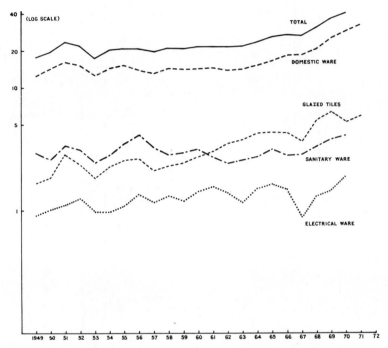

Figure 8.1   Pottery exports from the United Kingdom 1949–1971 (£ millions)

In Chapter 9 all aspects of industrial ceramics, including exports, are considered. Henceforth, in this chapter, we are concerned only with the exports of domestic pottery.

Table 8.1 shows the sources of imports of domestic pottery into the main importing countries for the year 1969. The figures are given in millions of United States dollars. The last column in the table shows the total value of imports of the various countries. It is immediately obvious that the United States is by far the largest importer ($133·3 million); the second largest is Canada ($30·9 million), followed by France and Germany ($24·2 million and $21·8 million). An interesting figure is the relatively small one for the United Kingdom of only $8 million. This figure is low because the British pottery industry supplies such a wide range of domestic and ornamental ware

Table 8.1 WORLD TRADE IN DOMESTIC POTTERY IN 1969 ($m.)

Importer	Sources of imports															
	U.K.	Australia	New Zealand	Canada	W. Ger.	Italy	S. Africa	U.S.A.	Japan	France	Neth.	Bel.-Lux.	Switz.	Sweden	Other	Total
U.K.					1·2	0·9			1·6	0·5	0·4				3·3	8·0
Australia	3·8		0·5		0·6	0·4			4·7						1·3	11·4
New Zealand	1·1	0·5													0·3	1·9
Canada	17·2	0·2			1·1	0·8		3·0	5·2	0·2	0·1				3·1	30·9
W. Germany	2·8					4·1			1·6	1·5	1·9	0·9	0·4	0·2	1·6	15·2
Italy	1·2				14·3				1·4	1·5	0·1	0·1	0·1		3·1	21·8
S. Africa*	1·9				0·5				2·3							4·7
U.S.A.	20·1			0·1	10·0	8·0			83·9	2·5	0·5		0·3	0·1	7·8	133·3
Japan	0·4				0·1										0·8	1·4
France	1·2				13·4	4·6			2·1		0·7	1·1			1·1	24·2
Netherlands	1·2				10·4	0·8			1·0	0·6		1·4			1·4	16·8
Belg.-Lux.	0·1				4·7	1·3			0·8	1·9	1·7				1·3	11·8
Switzerland	0·8				4·3	0·5			0·4	0·6	0·1			0·2	2·0	8·9
Sweden	1·1				3·2	0·4			1·0		0·1				4·4	10·2
Totals†	60·1				80·8	25·7			137·1	12·7						300·5

Source: *United Nations' World Trade Annual* (1969), Vol. III.

*South Africa's imports are not available. Individual countries' export figures used.
†Including countries not listed above.

at such keen prices that foreign producers find it difficult to penetrate into the market. The tariff, which is only 3% *ad valorem*, could not be regarded as severely restricting imports. Finally, in Britain there is a marked preference for tableware and ornaments which are made by British firms.* The pottery industry has long had an impressive export performance; no less important in terms of the balance of payments has been its ability to cater consistently for the needs of the home market. Only Japan, among the major producing and exporting countries, imports less domestic pottery than Britain.

The bottom row in *Table 8.1* showing the total value of exports shows that the major exporter is Japan, followed by a rather poor second with Germany, and that Britain is third. The only other major exporter of domestic pottery is Italy.

The United Kingdom's column in *Table 8.1* shows the extremely high dependence of British pottery manufacturers on the United States and Canada. However, there is not much choice involved as the United States and Canada are the two major importers. Looking at Germany's column we find that Italy, France and the Netherlands and other European countries constitute Germany's main markets, Germany's sales to the United States being only half the sales there of the United Kingdom. Italy too, sells to European countries; nevertheless the United States is also Italy's main market. Finally, a glance at Japan's column reveals that, in 1969, Japan sold four times the value of the United Kingdom's exports to the United States ($83·9 million). The value of Japan's sales to the United States is, by far, the most impressive aspect of world trade in domestic pottery.

The exports of the domestic sector of the British pottery industry in 1969 were valued at £25 856 400 which is 45% of total output. It would seem that in the post-war years export markets grew faster than did the domestic market; the percentage exported in 1963 was 41% and in 1958, 42%. Pre-war, it was 33% in 1930 and 37% in 1924. What is important is the marked dependence of the domestic sector of the industry on export markets both now and in the past; there is a tradition of exporting on a substantial scale which stretches back to the eighteenth century. Why this should be so depends on many factors. Presumably emigration from Britain in the nineteenth century

---

* In 1968, 800 British housewives who had been randomly selected were asked: 'Have you ever purchased foreign tableware, ornamental ware or kitchenware?' Only 14% of the housewives stated that they had ever purchased foreign pottery. (The question was 'have you ever . . .', not 'have you . . . in the last year?') P. W. Gay and R. L. Smyth, *The Household Demand for Tableware in Great Britain in 1968*, University of Keele, Department of Economics, 1971.

laid a foundation for sustained external demand for traditional British pottery in the United States and in the Commonwealth. The uniqueness of British bone china was important as was the desire and ability of British pottery manufacturers to sustain and expand export markets. Obviously British pottery must have had and still has unique qualities which could not and cannot be reproduced abroad, outside of Japan, at comparable cost. In *Table 8.2* an attempt is made to com-

Table 8.2 THE UNITED KINGDOM'S SHARE OF THE WORLD MARKET* FOR SELECTED CONSUMER DURABLES (%)

Item	1953	1966	1969
Travel goods, handbags, etc.	18	5	5
Furniture and fittings	24	10	7
Footwear	35	8	8
Glasswear	9	9	9
Clothing (except fur)	24	7	9
Floor covering and tapestries	37	10	16
Domestic pottery	33	22	18

Sources: 1953 *United Nations' Handbook of International Trade Statistics*; 1966 and 1969 *United Nations' World Trade Annual* (SITC sections 831, 821, 851, 665, 841, 657 and 666).
*'World Market' is here defined as total exports by the United States, Canada, the E.E.C., Sweden, Japan and the United Kingdom.

pare the export performance of British pottery manufacturers with those of other industries which produce consumer durables. The table shows that in 1969 domestic pottery exported from the United Kingdom accounted for 18% of the world market, which was a higher share than any of the other product groups listed. The table suggests that many other comparable British industries have failed to sustain their exports to the same extent as the domestic sector of the pottery industry.

Figure 8.2 shows the value of exports of domestic pottery in millions of United States dollars and at current prices, from 1952 to 1970 by Japan, Germany and the United Kingdom. We have already noted that Japan and Germany exported more domestic pottery than the United Kingdom in 1969; the figure also shows that Japan pulled rapidly away from both the United Kingdom and Germany after 1955 and the close similarity of Germany's and the United Kingdom's performance over the period, with Germany overtaking the United Kingdom in 1960 and keeping ahead thereafter. The contrast between Japan and the United Kingdom suggests that the export performance of British pottery manufacturers since 1955 should have been better. The value of British domestic pottery exports rose by 43% between

1955 and 1969. Export prices probably rose by more than this (the wholesale price index of domestic pottery rose by 66% over the period) so that the *volume* of exports probably fell.

It is clear that the British pottery industry failed to grasp a golden opportunity to expand in the world market. There are many reasons for this—labour shortages, problems of reconstruction, lack of finance, a lack of sufficient aggression in marketing—all these problems, however, were overcome by Japan, as the figures show. In Japan the pottery industry was helped considerably by government loans and subsidies and by massive aid from the banks. In Britain the Govern-

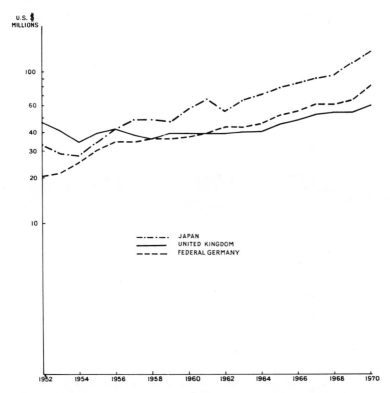

Figure 8.2 Exports of domestic pottery: Japan, Germany and the United Kingdom 1952–1970 (SITC 666)

ment helped the pottery industry with reconstruction in the immediate post-war years and after that it was left alone to put its own house in order. The firms in the pottery industry wanted it that way, and the outcome was that nearly all the growth in world markets

Table 8.3 EXPORTS OF DOMESTIC POTTERY AND OF MANUFACTURED GOODS 1955–1969 (INDEX NUMBERS OF VALUE)

Year	(i) World Manufactured goods	(ii) World Pottery	(iii) U.K. Manufactured goods	(iv) U.K. Pottery	(v) World (ii) as % of (i)	(vi) U.K. (iv) as % of (iii)
1955	100	100	100	100	100	100
1956	115	102	103	92	89	89
1957	126	102	114	86	81	75
1958	122	104	124	95	85	77
1959	131	114	130	94	87	72
1960	150	123	127	94	82	74
1961	159	117	133	94	74	71
1962	167	126	142	97	75	68
1963	181	133	149	97	73	65
1964	206	149	151	107	72	71
1965	232	159	180	115	69	64
1966	284	174	193	126	61	65
1967	277	177	189	126	64	67
1968	319	196	202	124	61	61
1969	375	235	234	143	63	61

for domestic pottery was catered for, admirably, by Japan and not by the United Kingdom. The weakness lay in the British economy rather than in the pottery industry, which is merely one very small part of the whole. Probably no economy at any time ever expanded its exports as fast as did Japan since 1955 and pottery was one item in the phenomenal Japanese export drive. This point-of-view is supported by the data contained in *Table 8.3*. The table shows that world exports

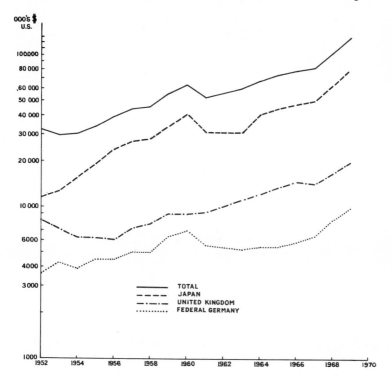

Figure 8.3   United States' imports of domestic pottery 1952–1969 (SITC 666)

of manufactured goods increased by 275% between 1955 and 1969 whereas 'world'* exports of domestic pottery increased by 135% over the same period. In contrast the exports of manufactured goods by the United Kingdom increased by only 134% between 1955 and 1969 and domestic pottery exports increased by 43%. Columns (v) and (vi) of *Table 8.3* show the growth of pottery exports from the United Kingdom relative to the growth of its manufactured

---

*'World' exports of domestic pottery defined as the exports of Japan, Germany and the United Kingdom.

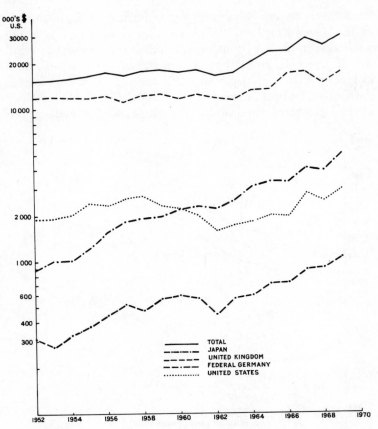

Figure 8.4 Canada's imports of domestic pottery 1952–1969 (SITC 666)

goods exports has been similar to the growth of world exports of pottery relative to the growth in the world of exports of other countries' manufactured goods. Over the period the world ratio fell from 100 to 63 and the United Kingdom fell from 100 to 61. Obviously the British pottery industry has suffered from being part of a slow-growth economy.

Figures 8.3, 8.4 and *8.5* show how Japan, Germany and the United Kingdom fared over the period 1952–1969 in the three main markets for domestic pottery, the United States, Canada and Australia.

ORGANISATION

Pottery firms normally appoint import agents to sell their wares in the various markets abroad. In large markets, such as the United

States, a number of agents will be appointed, one or more for each main region. The agents tend to be wholesalers who handle a wide range of hardware items, including pottery and glass. Agents usually

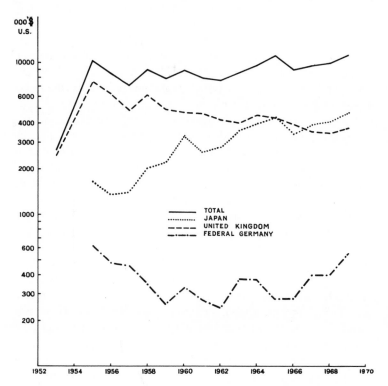

Figure 8.5   Australia's imports of domestic pottery 1952–1969 (SITC 666)

pay for the goods when they are imported, rather than work on commission, and if ware is also sold directly to retailers then the importer will be given a wholesale discount. It is usual for the same prices to be charged to wholesalers abroad as at home for tableware.* Since the early 1920s an earthenware price list has been circulated to overseas

*R. A. Cooper, K. Hartley and C. R. M. Harvey, *Export Performance and the Pressure of Demand*, Allen & Unwin, London (1970). See 'The U.K. Pottery Industry', chapter 6, page 110: 'There seemed at times to be an almost moral feeling against differential pricing. Granted the constraints of anti-dumping laws, the comfortable demand conditions and many firms' high export to total sales ratio, such inflexibility on pricing in *all* markets was surprising. It was equally remarkable that the possibility of higher prices for exports as compared with home sales was not mentioned, given the general situation of buoyant export demand and capacity limitations.'

buyers and exporters usually quoted their prices from the list. Usually a price for a 10 or 12 inch plate was quoted and all other prices in a set were related to it. According to the quality of the body and the decoration and the standards of selection used a price would be high or low, denoting high or low costs of production. Given that the price list was based on costings made many years previously, and that relative costs had changed over time, some items in tea, coffee and dinner sets yielded profits and others yielded losses. Items which have awkward shapes—jugs, sauce boats, tea and coffee pots, serving dishes, etc.—tended to be unprofitable and items that had come to be mass-produced—cups, saucers and plates—yielded profits. Because of costing anomalies, manufacturers tended to sell 'stripped-down' sets containing only profitable items. This, of course, was not intended as the price list had been issued to simplify the task of foreign purchasers and to prevent price-cutting by agreeing that all British exporters should use the same list of prices. The price list was finally abandoned in 1970.

Many import agents have had long-standing associations with British manufacturers and provide excellent service. Others do very little to promote sales. The blame for an unsatisfactory agent must lie with the manufacturer; he should find a better agent or pay regular visits to agents to stimulate them into action. Pottery firms attempt to arrange for their directors to visit their export markets on a regular basis. Experience has shown that this is the best way, as in other industries, to sustain orders and establish proper working arrangements with agents. Many pottery manufacturers have appointed agents in countries scattered all over the world so that it is virtually impossible to visit more than a few in any one year. Perhaps small and medium-size firms would do better if they concentrated their selling efforts in fewer markets. However, because any one export market or even a group of markets may be closed down suddenly by the imposition of quotas or excessively high tariffs, most manufacturers prefer to maintain contacts with as many markets as possible. Overseas buyers, importers, wholesalers and retailers frequently visit Britain, and pottery firms maintain showrooms in London and at their factories for their convenience.

It is rare for a manufacturer to believe that an agent is doing as good a job as he would do himself. Markets are normally developed by agents; then, if the scale of operations justify it, the exporter will establish his own warehouses, showrooms and salesmen in particular foreign markets. This can be done only by large firms with substantial sales in particular countries. We find that overseas branches of British pottery firms that act as importers are concentrated in New York and

Toronto. Operating overseas branches is a substantial advantage in promoting exports and it is one that is denied to small businesses. A really good import agent could be just as good, but they are difficult to find, retain and trust. The desire to establish and/or expand overseas subsidiaries which would operate showrooms, maintain close contacts with retailers, supply from stock and undertake market research was an important consideration in some takeovers. Only large firms can sustain the volume and variety of output needed to render properly staffed overseas branches profitably. Of course, firms such as Royal Worcester, Wedgwood and Doulton/Allied English still operate by means of import agents in many of their overseas markets.

The organisation of exporting varies considerably according to the countries, the supplying firms and the type of ware involved. Small firms may operate through export agents who specialise in particular areas—Europe, South America, the Far East, etc.—and all firms obtain orders by means of participating in trade fairs or by direct contacts with retailers and wholesalers. Rather than attempt to generalise more about all markets, we will be specific about the United States market. Extracts are reproduced from an article which appeared in 1969 in the *Pottery Gazette and Glass Trade Review* (now *Tableware International*) on 'The Importing Scene'. It serves as a reminder that exporters sell status symbols, good taste and dream fulfilment, not merely dinner sets and rose bowls; and that successful exporting is based on scrupulous attention to detail.

It (the United States) is a good and vast market. It has the economic means to purchase ware. But the ware must be right. . . . The average bride in the United States is 17·9 years and, it is estimated, accounts for 60 to 70% of domestic pottery sales. Most of the advertising, in a publicity-conscious country, is aimed at influencing the bride-to-be to register her wedding gift list at one of the big stores or shops. Thus most importers take space in the bride magazines, and perhaps in weeklies such as the 'New Yorker' and the 'New York Times' magazine. The bride is young, has lately left college and is still very much under the influence of her home economics teacher. Because she is unsure of herself, she chooses mainly safe patterns, such as the pale green-on-white 'Allegro' of Royal Worcester, or the palest shadow of a pattern, white on white. Royal Doulton's 'Clarendon', a narrow, pale green and gold border, is popular with this category at present, too. Bridal registers are common in most shops and stores. There is a move to sell stronger patterns to her—Wedgwood's yellow 'Mimosa', for instance, although this company still names pale olive 'Westbury' as top seller. All the leading importers carry out surveys into taste. They lecture at clubs throughout the States, hold competitions at women's colleges, and also set out to educate the home economics teachers, one stage back in the influence stakes. . . . But the British subsidiaries are increasingly aware of a new potential customer—the mature woman with a young outlook, who, as a successful executive's wife, may decide that the time has come to buy a new dinner service. For her, perhaps the popular 'Blue Regency' pattern of Worcester, with gold border and cobalt band. . . . Bone china tea-cups and saucers, given as engagement presents, are a good sale, and the *demi-tasse* coffee is probably more popular in the United States than here. Some years ago only the open cup could sell, but taller cups are now well accepted. One *sine qua non* for efficient sales on the American market is that patterns should be dishwasher-proof, and salesmen who wish to sell must show that ware can safely be put into the dishwasher, quite commonly owned in that high standard of living country. Wedgwood now advertise their ware to the effect that it is recommended for use in dishwashers, provided consumers use the detergent which has been found safe. Competition is strong from Lenox, an American concern described as the largest in the world,* claiming about 49% of the United States bridal market. This they capture mainly by advertising. They also run an annual competition for the young bride, which no doubt gains many converts for Lenox. Lenox also produce glass and bone china under the name 'Oxford'. Price is not a ruling factor in this rich market. It is a constant battle for shelf space. To help fill them each company has regional salesmen on the road, the largest, more than a dozen. They visit retail outlets constantly. The central buying habit is gaining in the United States, and there are six to eight strong buying groups. In New York, probably the best known stores for tableware are Altmans, the big department for dinnerware; sensible, well-stocked Bloomingdales; smart, modern Macy's and, of course, select Tiffanys. There are several stores in other States of America which enjoy nation-wide fame, Marshall Field's of Chicago, for instance, Nieman Marcus of Dallas, Texas; Rich's (Atlanta), Stix, Baer, and Fuller (St. Louis) and so on. There are specialist stores such as Gump's in rollicking and beautiful San Francisco, and Jensens in New York, where Wedgwood have a section.

*So described before the formation in 1972 of Royal Doulton Tableware Ltd.

Following upon devaluation of the pound sterling by 14·3% on 18 November 1967, domestic pottery sales increased substantially. The increases were not caused only by the devaluation, however, as the industry would probably have benefited anyhow from the upward surge in world trade in manufactured goods. Also, at the time of devaluation many pottery groups had finally succeeded in obtaining capacity output in really long runs from their reconstructed factories. An interesting reaction to devaluation, which we discovered by means of a questionnaire, was that all firms increased their prices to offset or almost offset the effect of devaluation on tableware prices in foreign markets. By November 1967, price increases were overdue, so firms used the opportunity to earn extra profits and use them to expand expenditure on advertising and sales promotions in foreign markets. Also firms aimed, after they had increased prices, to guarantee customers steady prices for a few years to come. As Japanese manufacturers were finding it difficult to hold their prices this was good competitive strategy. The way domestic ware manufacturers reacted to devaluation suggests that they had become more sophisticated in their selling methods in the 1960s than in previous decades.

EXPORT PERFORMANCE AND THE PRESSURE OF DEMAND

A research project at the University of York was concerned with 'some of the factors affecting the export performance of firms in a limited number of industries during the period 1958–1966'. The main concern of the project was to test the hypothesis that 'in the short run, variations in exports are a function of the pressure of domestic demand'. The results were published in 1970 in a book: *Export Performance and the Pressure of Demand, A Study of Firms,** chapter 6 of which is devoted to the pottery industry, one of the four selected industries. Their investigation, based on 'interviews with senior managements, and econometric analyses of the performance of individual firms and of the industry as a whole . . . provided little or no support for the hypothesis that domestic pottery exports are adversely affected by high pressure of demand and benefited by relatively low pressure of demand'. This confirms what we too had been told by pottery managers. They normally treat the home market as one market out of many and give priority to obtaining and fulfilling overseas orders. It simply

*R. A. Cooper, K. Hartley and C. R. M. Harvey, *op. cit.*, Allen & Unwin, London (1970).

is not possible to sell more abroad, it is argued, if there is a recession on the home market, unless a firm is well-established in overseas markets. Also, if orders are lost in one overseas market it may be easier to dispose of it in another one than sell it on the home market. Many shapes and patterns sold overseas are as unlike those for the home market as chalk is unlike cheese, and switching output between markets is simply not considered. In Chapter 7 the limitations of the home market are stated and some reasons for the limitations are presented. There appear to be possibilities of expanding the home market considerably in the long-run. These possibilities are extremely important because of possible adverse effects following upon Britain's entry into the Common Market. There does not appear to be scope for an expansion of British exports of domestic pottery outside of North America and Europe in the face of strong competition from Japan and Germany.

ENTRY INTO EUROPE

Pottery manufacturers, in common with most other British businessmen, have long regarded entry into Europe as a good thing from the point of view of their own firms and the country as a whole. In spite of the low tariff of 3% imposed on tableware, imports into Britain have been modest. The abolition of the tariff is not, therefore, expected to make much difference. In contrast, members of the E.E.C. enjoy substantial protection against British pottery (16·2% in 1971 and 12·6% in 1972). Entry to the E.E.C. is therefore expected to give rise to a substantial increase in exports to Europe. Pottery manufacturers, particularly those which specialise in fine china, are concerned about their extremely high dependence on the North American markets. The most obvious way to remedy the situation would be to substantially increase exports to Europe. Entry into Europe offers this attractive possibility. It also offers the less attractive possibility that European manufacturers will succeed in penetrating deeply into the British market. This assumes that in the 1960s the European producers did not pay much attention to the U.K. market. Finally, the gains in Europe must be sufficient to offset reductions in exports to Commonwealth and other important and traditional markets. In 1971 the margin of Commonwealth preference was 20% in Canada, 10% in Australia and 40% in New Zealand. Reductions in exports to non-E.E.C. markets must occur because most countries will increase or impose for the first time tariffs on imports from Britain after Britain has entered the E.E.C. We must not overlook the fact that as the Common Market creates new export opportunities it will also curtail old ones.

THE DOLLAR

The imposition, in August 1971, by President Nixon of a temporary 10% tax on dutiable imports spread gloom and despondency among pottery manufacturers. The fact that the tax was aimed primarily at Japan was small comfort. The weakness of the dollar had forced the United States to act to restrict imports and thereby strengthen its balance of payments. Fine china was particularly vulnerable as it obviously is not one of the basic necessities of life. The decision which was finally taken, in December 1971, by the United States to devalue the dollar and increase the price of gold appeared to offer prospects of stable conditions for world trade in the 1970s. Quickly despondency gave way to optimism in North Staffordshire. The episode vividly illustrates the uncertainties which face an industry which is heavily dependent on exports. What else can tableware manufacturers do but cross their fingers and hope that the United States market will continue to expand and accept imports in substantial quantities?

Part III

TILES, SANITARY WARE, ELECTRICAL WARE
AND INDUSTRIAL CERAMICS

9

INDUSTRIAL CERAMICS

A number of domestic ware manufacturers object strongly to the use of the words 'the pottery industry'. They explain that they are manufacturers of ceramics and a word that was good enough for the Ancient Greeks is good enough for them. We have persisted with old-fashioned nomenclature in this book for the domestic sector of the industry, mainly because everyone knows about pottery whereas a few are puzzled by ceramics. However, for this chapter industrial ceramic looks and sounds better than industrial pottery.

This chapter has been divided into four sections: Tiles, Sanitary Ware, Electrical Porcelain and Industrial Ceramics. Under each heading the major firms involved are described and export performance and channels of distribution on the home market and other matters are discussed. The brief description of the manufacture of large insulators 'for modern high-voltage apparatus and sub-station equipment' is based largely on the excellent pamphlet which contains a sequence of 15 coloured photographs, *The Production of Taylor Tunnicliff Insulators* by Taylor Tunnicliff. The testing of large insulators is described and illustrated in a pamphlet *Making Porcelain Insulators* by Doulton Industrial Porcelains Ltd.

TILES

According to the theory of competition of Jack Downie,* if one firm

*Jack Downie, *The Competitive Process*, Duckworth, London (1958).

among many makes an innovation which lowers its cost significantly below the costs of rival firms, then that firm will grow rapidly at the expense of its rivals and they will either go out of business or be taken-over by the innovating firm. The outcome of competition will be monopoly, or oligopoly (a few firms), if one or two of the rival firms survive and succeed in making worthwhile innovations too. The tiles sector of the pottery industry admirably fits this theory. Because tiles are fairly standard articles, as compared with tableware and ornaments, competition between firms is intense and there is little protection against a firm that obtains a cost advantage and decides to use it. In the *Census of Production*, 1935, 58 establishments were reported as producing 'floor tiles and glazed wall and hearth tiles'. In 1960, 13 companies were members of the Glazed and Floor Tile Home Trade Association. 'Of these, three were subsidiaries of other members. Five of the 13 members engaged primarily in the manufacture of wall tiles, four engaged primarily in the manufacture of fireplace tiles, three engaged exclusively in the manufacture of fireplace tiles, one engaged exclusively in the manufacture of floor tiles. All those primarily engaged in making wall tiles also made some fireplace tiles and two of them also made floor tiles. All those primarily engaged in making fireplace tiles also made some wall tiles and one of them also made floor tiles.'* This passage is a reminder that industrial structures tend to be rather complicated. The major firms engaged in tile making in 1960 were: Pilkington's Tiles Ltd., Campbell Tile Co. Ltd., Malkin Tiles (Burslem) Ltd., Richards Tiles Ltd., H. & R. Johnson Ltd., T. & R. Boote Ltd. and Carter Tiles Ltd. Over the course of the next eight years, all of the companies were merged into two rival groups.

Pilkington's Tiles Ltd. acquired Boulton Tileries & Co. Ltd. in 1949, and subsequently Carter Tiles Ltd.; the firm is located in Manchester. Pilkington's have competed with success against their larger rival, H. & R. Johnson–Richards Tiles Ltd.; in 1971 it was acquired by Thomas Tilling, a financial holding company with interests, among others, in builders' merchants. In 1964 H. & R. Johnson acquired an interest in Malkin Tiles who were finally bought-out in 1968. In 1965 Richards Tiles acquired the Campbell Tile Company; both were long-established and large and active companies. Finally, in May 1968, a merger was arranged between H. & R. Johnson and Richards Campbell Tiles Ltd. The new company adopted the name of H. & R. Johnson–Richards Tiles Ltd. and the Chairman was Derek H. Johnson.

*Restrictive Practices Court, *in re* Glazed and Floor Tile Home Trade Association's Agreement, (1961 No. 13 (E. & W.)), October and November 1963 and January 1964. Henceforth referred to as Restrictive Practices Court, 1963/64.

(Other members of the Johnson family made Johnson Brothers (Hanley) Ltd. the largest manufacturer of earthenware in Britain and they are now concerned in making Wedgwood bigger and better. Members of the Johnson family appear to have a facility for making good business decisions, and more important, being in positions which enable them to carry their projects through to successful conclusions.) In 1966 the sales of H. & R. Johnson Ltd. were £6 800 000 and in 1971 the sales by the new company had risen to £18 035 000. (In 1970 sales by Pilkington Tiles Ltd. were £6 271 000.) Why did the rapid changes in ownership which have transformed the tiles sector of the pottery industry happen? It is tempting to answer the question in terms of one enterprising and aggressive businessman, Mr. Derek H. Johnson, who seized on favourable technological and marketing opportunities. Mr. A. N. Smith, the Managing Director of Pilkington Tiles also contributed. In what follows our attention will be mainly directed towards the underlying economic changes which were successfully exploited by Pilkington's and the Johnson–Richards Group.

The Restrictive Practices Court's decision in 1964 to approve the Glazed and Floor Tile Home Trade Association's Agreement was an important decision. Only in the tiles sector of the pottery industry is it possible to fix prices legally. The Court, by its decision, made it legal and respectable to standardise tile sizes, to establish minimum prices, to control the sale of reject tiles and prevent price cutting. The 60-page report of the Court's proceedings provides a detailed description and appraisal of how the industry was organised and how tiles were made and marketed in 1960.

During World War II standardisation of tile sizes had been imposed on the industry by government control. Realisation 'of the benefits of standardisation' led them after the war to attain a massive degree of standardisation; in 1961 only a negligible proportion of their production was of non-standard tiles. The basic shape of their tiles was square and they concentrated production on two standard sizes for each of their basic tiles (6 × 6 inches and $4\frac{1}{4}$ × $4\frac{1}{4}$ inches for wall tiles and 6 × 6 inches and 4 × 4 inches for fireplace and floor tiles), and also on a considerable range of ancillary types, sizes and fittings.

By their agreement members agreed to sell their standard tiles (including the ancillary types, sizes and fittings) at the prices set out in the association's price list, and to sell all other tiles (non-standard tiles) at a price which was not less than 25% in excess of the price of the most comparable standard tile or at the actual cost of manufacture, whichever was the greater. Members were forbidden to sell tiles below standard quality. Those had to be offered to the association's clearing house, which was run by a manager, appointed by the association, who had a complete discretion as to the terms on which sub-standard tiles were sold. The profits and losses of the clearing house were apportioned between members in proportion to their disposals to it. It had to take all sub-standard tiles offered by members. The association's prices were delivered prices, and there were quantity discounts, a trade discount of 20% and a cash discount of 5%. There was a surcharge of 5% for deliveries of floor tiles and of 10% for deliveries of slabbed fireplace goods to Scotland, Northern Ireland, the Isle of Man and Isle of Wight. There were corresponding surcharges of $7\frac{1}{2}\%$ and $12\frac{1}{2}\%$ for the Channel Islands.

Members' voting rights in respect of prices were related to their turnover in the home market in the preceding three years. Prices generally could be increased only if members who had 80% of voting rights approved, and prices in respect of any particular class of goods could similarly be increased only if members having 80% of the voting rights in that class approved. Prices could normally be reduced by any two members who had 25% of the relevant voting rights (or, in the case of fireplace enamels, 5%). There were complicated provisions by which one member having only 2% of the relevant voting rights, or two members having 20%, $12\frac{1}{2}\%$ or 5%, could have prices reduced where there had been specified decreases in the aggregate turnover of all members in the relevant class of goods, or in the utilisation of their aggregate capacity in the relevant class. In December 1962, a new provision for price reduction was introduced, whereby any member who had perceptibly lowered his costs by some innovation, change in process or improvement in efficiency, could reduce prices of the relevant goods for six months.

'There was a high degree of co-operation and mutual assistance between members as to their technical and commercial problems. They exchanged technical information; eight members regularly circularised details of their process costs to all members, any member's technical executive could freely communicate with another's and seek his advice, and the doors of members' works were freely open to each other. Their free and open exchange of information had assisted their efficiency. The industry was efficient and progressive, members' profits were no more than reasonable, and post-war price increases had been very moderate.'*

An important aspect of the agreement is that long runs in a few

*Restrictive Practices Court, 1963/64 op. cit. pp. 240–241.

sizes result in substantial cost saving. The tiles are manufactured from fairly dry powder which is fed into pressing machines which stamp

Plate 9.1 Glaze storage tanks (H. & R. Johnson–Richards Tiles Ltd.)

Plate 9.2 Dipping house conveyor system (H. & R. Johnson–Richards Tiles Ltd.)

out tiles within very fine limits of accuracy. Compared with domestic ware and sanitary ware, the forming process is greatly simplified and drying is not a problem. (Attempts have been made to produce plates by a similar way. Unless the ware is entirely flat it is difficult to avoid weaknesses. The British Ceramic Research Association developed a plate-making machine which used dry dust as its raw material. It was found that the cost per plate was unduly high, in spite of the seeming simplicity of the making operation, and the machine was not developed commercially.) The tiles are made by presses which can be grouped and controlled by one operator, the tiles are subsequently scraped and glazed automatically and decorating is also automatic, and the loading and unloading of the kilns has been greatly simplified.

A number of elements were brought together which transformed both the method (and cost) of making tiles, and the markets for them. Much of the pioneering work was undertaken by H. & R. Johnson, and this helps to explain why H. & R. Johnson–Richards Tiles now dominate their sector of the industry. The first major break-through was achieved when limestone was added to china clay in the body composition of the tiles. This enabled tiles to be made with accurate dimensions. Pre-1955 tiles can be identified by the fact that each tile approximates to the standard size. Size variation, of course, prevents tile fixing being something that anyone can do. Subsequently it was discovered that satisfactory tiles could be manufactured to a reduced thickness of $\frac{5}{32}$ in. instead of $\frac{3}{8}$ in. This innovation reduced firing costs and transport and storage costs; in addition it rendered simpler and more attractive packaging possible, and it made it easier to attach tiles to walls.* With accurate sizes, virtually anyone with the inclination could read simple instructions and fix tiles. A new market quickly developed which was supplied by 'do-it-yourself' shops. Tiles were also produced with spacing lugs on the back; when the lugs touched, the tiles were the right space apart. Without lugs the work of a professional fixer could be seen to be superior to that of an amateur; with them the amateur's work could look as good as that of a professional. In 1957 Richards Tiles developed new adhesives which eliminated the need to mix dry ingredients with water. Subsequently the tile manufacturers formed a company to market tile adhesives. A most important innovation was the development of standardised colours. In consequence, colours could be matched from stock in all parts of the country. Before it was found possible to standardise

*The search for thinner tiles had been stimulated by the tariff of Venezuela being calculated by weight instead of area. This is a rare example of a tariff benefiting the supplying country.

colours it was necessary to manufacture tiles in batches. To avoid colour variations, builders' merchants bought complete batches, or else used only cream or white tiles. This greatly increased the cost of holding stocks and rendered re-ordering difficult. It also meant complications for the makers, as there was always uncertainty about the proportion of each firing that would be perfect. Manufacturers therefore tended to produce more than had been ordered of each size

Plate 9.3 Selecting glazed tiles (H. & R. Johnson–Richards Tiles Ltd.)

and colour, in case kiln loss would be excessive. The consequence was that factories, warehouses and yards were littered with small quantities of perfectly good but unwanted tiles. All this was eliminated once proper colour control had been achieved. The market was also expanded by improving the colours and textures of tiles, in particular by using contemporary abstract designs. Standardisation meant that meaningless variations in sizes were eliminated. It did not, however, reduce variety in colour, patterns or textures.

H. & R. Johnson–Richards report that a square yard of their tiles were priced higher in 1919 than in 1972. Admittedly the 1972 tiles were thinner, but given that it is surface quality that is important this should not be regarded as product deterioration over the period. In contrast, the cost of hiring a workman to fix tiles increased substantially over the same period. It is the divergence of the two prices which helps to explain the success of do-it-yourself in the 1960s as far as tiles are concerned. It is not easy to think of any other product that was sold for less in 1972 than in 1919. The reduction in price indicates how tile production is now semi-automated.

It is the relatively low prices of ceramic tiles which have enabled the tile manufacturers to compete so successfully with substitute materials whether they be paints or wall papers.

Tile manufacturers offered substantial quantity discounts to customers who placed really substantial orders. A consequence of their pricing policy was that specialist tile distributors appeared to serve builders' merchants and various retail outlets. The tile manufacturers ceased to do their own wholesaling. Instead, attention was paid to

Plate 9.4   Central storage warehouse (H. & R. Johnson–Richards Tiles Ltd.)

providing prompt delivery and convenient packaging. The growth of the market was also promoted by close contacts being maintained with architects. They decide how new walls will be surfaced. All aspects of public relations were used by the tile firms to stimulate a public awareness of their products and to increase their use on public buildings as well as in homes.

Figure 9.1 shows an index of house-building and renewal activity between 1946 and 1969, and an index of the volume of tiles produced as measured in square yards. It may be seen from this index that the output of tiles kept pace with the building boom until the early 1960s.

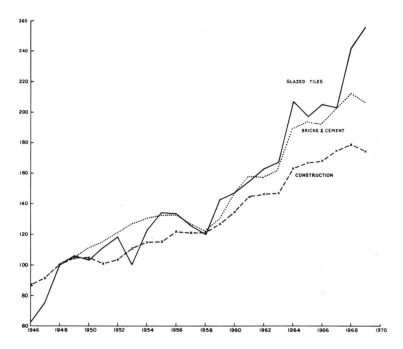

Figure 9.1   Production of glazed tiles, bricks and cement, and housebuilding: Index numbers of output 1948–1969 (1948 = 100)

What is remarkable is that as building activity slackened the output of tiles kept on expanding. Only to a limited extent is this explained by exports. The explanation lies in the new market that was developed by 'do-it-yourself' enthusiasts who used tiles to improve old property.

The firm of H. & R. Johnson–Richards Tiles Ltd. (in the early years H. & R. Johnson Ltd.) was responsible for a number of the innovations in production and marketing that were introduced. The firm's Chairman, Derek H. Johnson, was a powerful advocate of co-operation between firms to avoid price cutting and to share the costs of developing new machines and processes. (The willingness of tile manufacturers to co-operate contrasts markedly with the 'go-it-alone' mentality which tended to dominate the other sectors of the industry.) Johnson built up a team of fitters to develop and install new machines and equipment. The machines and equipment were new only in the sense that well-tried mechanical methods in other industries were being applied for the first time in the pottery industry. It could not be claimed that a high degree of engineering skill was involved. A flow of materials and products through factories, kilns and warehouses was achieved which minimised labour requirements. Overhead conveyors, in particular, reduced the number of workers required to fetch, stack and carry tiles. The mechanisation process caused few redundancies as the labour force was reduced by not replacing natural wastage. A system of regular consultations between managers and workers' representatives in the factories appears to have developed more rapidly in tiles than elsewhere in the industry. The merger between Johnson and Richards quickly produced results. Production was concentrated in the best plants and old ones were eliminated. A new corporate image quickly appeared, and the sales forces of the two firms were amalgamated. The contrast is marked with some of the mergers in the industry where rationalisation was achieved, if at all, only gradually.

The energies of the Johnson–Richards management team were not restricted to the British Isles. A number of overseas subsidiaries were established. A factory was established in South Africa in 1955. Further factories were established in Canada (1957), India (1958), Australia (1964) and New Zealand (1968). There is a Johnson–Richards marketing company in New York and one was opened in Sweden to gain experience of selling tiles in a European country. Further developments in European countries are contemplated. The establishment of factories abroad is a fairly normal reaction by a large company to tariffs having been imposed by countries which had, before protection was introduced, imported in substantial quantities. It is the policy of Johnson–Richards to standardise their tile sizes, trade names, packaging materials, etc. between countries, so that the output of one country can be substituted for that of another. In addition, it is possible that there are long-run advantages to be gained by being an international company.

Between 1958 and 1968 the profits earned by H. & R. Johnson rarely fell below 20% on capital (see *Table 9.1*), and they supplied the means to expand. It is unlikely that any other large firm in the pottery industry sustained such a high rate of profit for so long. In 1969 and in 1970, however, these profits fell to 11% and then to 8%. Profitability was checked by a slackening of demand and costs being adversely affected by snags developing in the Group's reconstruction

Table 9.1 H. & R. JOHNSON–RICHARDS TILES LTD. AND PILKINGTON'S TILES LTD.: PROFITS, 1955–1971

Year	Johnson–Richards			Pilkingtons		
	Capital employed	Pre-tax profits	Profits as a % of capital employed	Capital employed	Pre-tax profits	Profits as a % of capital employed
1955	995	357	35·9	982	101	10·3
1956	1 278	400	31·3	1 147	−1	—
1957	1 429	422	29·5	1 071	−80	—
1958	1 745	445	25·5	1 053	46	4·4
1959	2 061	634	30·8	1 047	190	18·1
1960	2 251	543	24·1	1 080	165	15·3
1961	2 639	711	26·9	978	162	16·6
1962	2 913	712	24·4	1 060	162	15·3
1963	3 086	526	17·0	1 266	195	15·4
1964	3 904	743	19·0	1 647	247	15·0
1965	5 005	983	19·6	3 469	416	12·0
1966	5 159	1 154	22·4	3 882	314	8·1
1967	5 613	1 038	18·5	3 887	329	8·5
1968	6 542	1 301	19·9	3 986	414	10·4
1969	12 389	1 347	10·9	3 974	570	14·3
1970	12 702	1 029	8·1	4 463	611	13·7
1971	13 252	1 637	12·4	—	—	—

Source: Moodies Services Ltd.

programme. As Johnson–Richards were experiencing difficulties, albeit only temporarily, their main rivals, Pilkington Tiles Ltd., increased their profit rate to 14% in 1969 and 1970. Part of this increase may, however, have been contributed by Poole Pottery Ltd., a subsidiary company of Pilkington. Anthony Hilton reported on the transformation of Pilkington Tiles in *The Observer* on 1 February 1970 as follows:

The Pilkington recovery owes a lot to its 43 year-old Managing Director, Arnold Smith. After research at Cambridge, where he read physics and maths, Smith moved into industry in the Potteries and graduated from research to technical work, and then into production. In 1956 he moved to Manchester and joined Pilkington, then a tightly controlled tile manufacturer, founded in 1891, not to be confused with the giant privately owned glass manufacturer, down the road in St. Helens. He joined the board the following year and became Managing Director in 1962. The Group's troubles began in earnest in 1964 when Pilkington merged with Carter, a pottery and tile manufacturer based in Poole, Dorset, and by the end of 1965 it was obvious that something had to be done. Smith launched his desperately needed rationalisation programme in the teeth of the slump in home building which followed the 1964 boom. But he pressed on and pushed through extensive production changes, product rationalisation and new sales techniques. 'Rationalisation takes a long time to bear fruit in this industry', he says.

We may regard the acquisition of Pilkington's by Tilling in 1971 as guaranteeing the long-run survival and success of the Company.

A third important supplier of tiles is Polycell Products Ltd., which supplies ceramic tiles mainly through 'do-it-yourself' shops. For a while Polycell purchased tiles from established tile manufacturers before manufacturing their own tiles. Their tiles are manufactured in sheets before being cut to size, which contrasts with the one-at-a-time technique used by most other tile manufacturers. Polycell Products Ltd. is a subsidiary of Reed International Ltd.

We conclude by observing that the tiles sector of the pottery industry has been unique in recent years in two important respects: (1) Making is a machine process and most workers in tile factories are machine minders. To date, the nature of their products has prevented such a high degree of mechanisation from being achieved in tableware or sanitary ware production. (2) Higher profits have been earned in tiles than in other sectors of the industry. They did not just happen: they were earned as a result of a series of successful innovations which provided consumers with good value for their money. The outcome has been a high degree of concentration with output largely in the hands of two firms. However, competition remains fierce and there is always potential competition from non-ceramic materials.

SANITARY WARE

A pedestal water closet, a wash basin, a bidet or even a urinal, on the factory floor, are beautiful objects, like finely-finished pieces of sculpture, in their own right. They are made of vitrified earthenware by highly-skilled craftsmen. Large pieces of perfect white, primrose, pink or turquoise sanitary ware demand a high degree of potting competence; they are objects to be admired if not cherished.

Some 80 years ago, ewers and basins and chamber pots were made in large quantities by the domestic sector of the industry. Once the 'trap' and 'siphon' had been perfected by close co-operation between sanitary engineers, plumbers and pottery manufacturers and miles of glazed stoneware pipes had been laid, a specialised branch of the pottery industry developed to produce sanitary fixtures and fittings. ('. . . It was not until the 1870s that Henry Doulton felt ready to expand this —the sanitary wares—side of the business. It was about this time that running water first became available upstairs as well as down, in the more expensive types of houses, enabling bathrooms and toilets to be installed on the upper floors, and later on, even wash-basins in the bedrooms.')*

Thomas Crapper patented his system of isolating or 'disconnecting' the living space from the sewer by means of a trap containing a water seal. The advances in plumbing and in sanitary fittings have enabled cities to grow seemingly without limit. The pottery industry must be credited with having provided cheaply and adequately some of the essential items for civilised living. Unfortunately, because of a lack of proper public regulation and provision, part of the price paid for proper sanitation in homes, offices, factories, etc. has been pollution of rivers, estuaries and lakes.

Sanitary ware factories are fairly widely dispersed throughout the country as compared with domestic ware factories which are concentrated in North Staffordshire. The cost of transporting sanitary ware products would appear to offset, for some firms, the advantages offered to pottery firms of location in Stoke-on-Trent. Originally location was on coalfields where there was clay and later clay was transported from Cornwall to where there was coal. Shanks chose a site in Glasgow and the firm still specialises in marine sanitary ware. At the other end of the country Johnson & Slater moved to Kent from the North of France as a consequence of World War I. Ideal–Standard is in Hull, a good location for an industrial process that requires strong-man labour, the sanitary ware factory being only one part of a heating and ventilating combine. Armitage, Ltd. has grown and prospered in Armitage, some 40 miles south of Stoke-on-Trent. Stoke-on-Trent benefited when Doulton purchased the Whieldon Sanitary Potteries there in 1937 and production was transferred from London. Other sanitary ware producers in Stoke-on-Trent include Twyfords Ltd. and Johnson Brothers (Hanley) Ltd. which, like Doulton, produce both sanitary ware and domestic ware in the City.

DEMAND AND DISTRIBUTION

Output of sanitary ware fittings is closely geared to building activity in the United Kingdom. Approximately one-fifth of output is ex-

*Desmond Eyles, *Royal Doulton 1815–1965*: Hutchinson; London, p. 109 (1965).

ported each year. In 1970, ignoring exports, the market was divided into three·parts: new housing demand (30%), replacement demand and demand generated by the renovation of existing houses (60%), and institutional demand (10%). The latter includes offices, factories, hotels, hospitals, schools, etc.

Since 1945, as a consequence of the high level of building activity, output has increased, accompanied by greater standardisation, improvements in production methods, and the extension and improvement of existing factories. Demand, however, has not been steady; the first severe recession occurred in 1956 and a recovery in 1971 had been preceded by five lean years. It was about 1970 before once-fired ware replaced the traditional twice-fired ware, this was an important technical innovation as it meant a greater throughput from existing kilns and a reduction in labour and other costs. Even now sanitary ware production in Britain is not highly mechanised and productivity improvements have been achieved by better methods of handling, standardisation of materials, improvements in moulds and drying processes. The introduction of the battery method of production, described below, into all sanitary ware factories in 1970–1971, may be regarded as the beginning of a substantial degree of mechanisation.

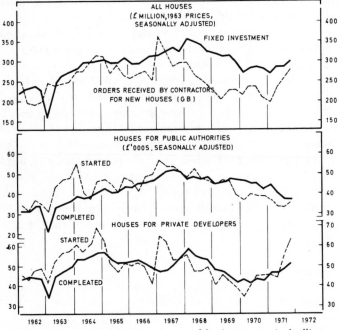

Figure 9.2　Forward-looking indicators of fixed investment in dwellings

After 1968, as may be seen from *Figure 9.2* fixed investment in dwellings fell sharply and orders received by contractors fell from 1967 onwards. *Figure 9.3* shows similar movements for the whole construc-

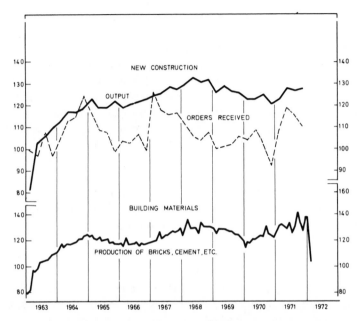

Figure 9.3   Investment in the entire construction industry (1963 = 100, seasonally adjusted)

tion industry and the lack of growth of demand for building materials, of which sanitary ware is one small part, from 1968 onwards. Sanitary ware producers have been helped considerably by more attention being paid by local authorities to the renovation of old property in recent years. The manufacturers regarded the slackening of demand towards the end of the 1960s as merely a temporary phenomenon and, in common with many other British industries, they entered the 1970s with substantial excess capacity. It was possible to direct only a fraction of the orders lost on the home market to the export market and, in times of recession, output diverted to the export market yields very low profit margins. Since 1968 profits in ceramic sanitary ware have had to be earned by means of strict control of production costs and hard selling.

A peculiarity of the demand for sanitary fittings is that it is highly concentrated in the hands of architects and builders. Few house purchasers enquire about the type, quality or maker of the sanitary fittings; in contrast, they tend to take a keen interest in kitchen fittings which increases the demand for stainless steel and plastics but not ceramics. Sanitary fittings tend to be regarded as part of the building best ignored; this renders new and possibly improved fittings difficult to sell. Also if purchasers or tenants are not aware of a lack of luxury or extra elegance in fittings, then builders will install merely adequate ones. Thus life is made hard for sanitary ware manufacturers as profit margins on bread-and-butter lines are, of necessity, low, and lines which would be expected to yield good profit margins are difficult to sell. In the 1960s there were signs that attitudes were changing and some builders were offering new houses with a choice of basic or luxury bathrooms. The characteristics of the products offered a challenge to marketing managers and they have responded. They operate well-appointed showrooms and issue attractive catalogues and brochures, and they maintain close contact with architects, builders' merchants, plumbers' merchants and builders. Most important of all, close attention is paid to design, the product range and prompt delivery. The Council of British Ceramic Sanitaryware Manufacturers supplements the selling activities of the member firms by exposing inadequate sanitary facilities in hotels, airports, offices and shops, etc. and making the general public aware of the benefits of bidets, silent water closet suites and extra bathrooms and toilets. Apparently the public tend to be as attached to cracked and stained pedestals and wash basins as they are to similarly affected cups and plates.

Whereas in tableware there has been a marked shift from selling to wholesalers to selling direct, in sanitary ware sales are made almost entirely to plumbers' and builders' merchants. Even when a builder has a large order for fittings it will not be placed direct; delivery to his sites will be made by the merchant and not by the manufacturer. There are a number of reasons for this state of affairs. Firstly, the wholesaler may act as an assembler. Before delivering, say, washbasins, taps may be fitted and waste-pipes supplied. The sanitary ware manufacturers, however, usually operate their own units for producing brass and plastic fittings; nevertheless, merchants may prefer to add their own bought-in components. Secondly, the builders' merchants are responsible for providing a vast range of materials and components to widely scattered sites at particular times. Credit must be arranged and debts collected. These are tasks for specialists and not for manufacturers. Thirdly, ceramic sanitary fittings and fixtures constitute merely one sub-group of 'baths and sanitary equipment'. It must not be overlooked that merchants find it convenient to operate with 20 commodity sections which range from 'roofing materials and ironmongery' to 'bricks and clayware and decorating materials'. It would not be possible for manufacturers regularly to deliver items direct to building sites without becoming involved in the complexities of all other materials. Finally, the manufacturers produce lines in hundreds, whereas builders require the delivery of only a few items at a time. Someone must break bulk and merchants are obviously better equipped for the task than the manufacturers of only one component. Some manufacturers may complain that large contracts may not be placed direct because of the convention that builders purchase only from merchants. (If a manufacturer should sell direct then he would expect there to be a severe fall in orders from wholesalers.) However, the wholesalers vary their mark-ups from as low as 5% to the more usual 35%, according to the amount of service rendered. The mark-up on individual items sold to householders for replacement purposes tends to be considerably higher. It could be argued that competitive forces operate to keep mark-ups from rising unduly. Wholesalers, perhaps understandably, keep their stocks of sanitary ware fittings as low as possible and do not act to any significant extent as a buffer between the builders and the manufacturers. Mainly because of the bulky nature of the products, manufacturers too hold only small stocks so that short-run fluctuations in output are inevitable.

There have been numerous takeovers and amalgamations of builders' merchants in the past 20 years; so, in bargaining, the small number of manufacturers face a relatively small number of wholesalers' groups. It is to be expected that manufacturers will complain about wholesalers just as farmers complain about the weather. In sanitary ware complaints are surprisingly few, perhaps this is because the need for merchants is so clear-cut. In the past, particular merchants have helped firms over bad patches and *vice versa* and goodwill has been firmly established. Manufacturers, however, must complain about something, so they complain about each other.

COMPETITION

The contrast with tiles is marked. Tile manufacturers co-operate closely on standardisation, minimum prices and research and development; the sanitary ware manufacturers co-operate on very little. Competition between firms is unregulated and intense. This leads to low prices and low profits. It also leads to unnecessary expenditures on research and development. For example, to achieve a body that need be fired once and not twice, each firm made trials independently. It was known that a solution would be found, yet co-operative effort was avoided. The lack of published statistics on sales is merely one indication that the various firms are more powerful than their trade association. Price cutting is a regular occurrence. Pedestals and wash-basins are sold at particularly low prices in spite of the fact that adequate non-ceramic substitutes are not available. Sales would not suffer if prices were somewhat higher on mass-produced and fairly standardised items. One way to reduce the incidence of price cutting is to produce products which are different. A surprisingly wide range of products is offered by the sanitary ware manufacturers. This is partly caused by excessive competition, but most of the trouble stems from a lack of standardisation in buildings. In addition, each export market tends to have unique sets of regulations. One would expect that if the manufacturers had co-operated more actively, more worthwhile standardisation would have been achieved. It is a fact of social organisation that only in particularly favourable circumstances will unregulated competition achieve an optimum allocation of scarce resources and encourage new production methods and new products. What is usually required is a proper blend of competition and co-operation, not necessarily government intervention beyond establishing rules; in the production and marketing of sanitary ware co-operation has been unduly neglected: One consolation is that the building trade obtain excellent products in a wide variety of shapes and colours at particularly low prices.

PRODUCTION AND FIRMS

Before World War II, sanitary ware was made in a larger number of smaller factories than at present. Between 1935 and 1963 the *Census of Production* reveals a fall in the number of establishments from 30 to 17 (the production of stone-glazed pipes was included in 1935 but not in 1963). The ware was twice-fired earthenware and some of it was fired in intermittent kilns. A typical factory produced some 5000 pieces each week. The introduction of tunnel kilns increased the output of each factory substantially and some factories ceased production. A similar outcome followed the introduction of once-fired ware; it not only reduced unit costs, it also resulted in a vastly improved product. The recent introduction of the 'battery system' will probably result in production being concentrated in yet fewer and larger factories. The concentration of output into a small number of large firms may be explained by commercial, research and transport economies which arise when a number of factories are controlled by one firm. Takeovers may also occur because of excessive competition; firms which refuse to co-operate sometimes agree to amalgamate. In 1969, 91% of the market was divided between four companies roughly as follows:

Armitage Shanks Ltd.	32%
Ideal Standard Ltd.	23%
Twyfords Holdings Ltd.	21%
Doulton & Co. Ltd.	15%

Shanks Holdings Ltd., of Glasgow, acquired two well-established firms, J. & R. Howie Ltd. and George Howson & Sons Ltd., in 1964 and 1966. In turn, Shanks were acquired by Armitage Ltd. in 1969 and the new firm Armitage Shanks Ltd. was established. Ideal–Standard Ltd. of Hull acquired John Steventon & Son Ltd., a firm which had grown on a basis of particularly low prices for standardised items, and after a series of takeovers in 1968 of plastics and plumbing firms, it announced that it was the 'largest manufacturer of plastic and vitreous china sanitary ware appliances in the United Kingdom'. In 1971 Ideal–Standard contracted somewhat by selling its interests in plastics, etc. Meanwhile, the firm which, within living memory, had been the largest and probably the best known one in the industry, Twyfords Ltd., fell to third place. Doulton & Co., which is the fourth largest British producer of ceramic sanitary ware, acquired Johnson & Slater Ltd. in 1968 after an offer had been made for the Company by Armitage Ware Ltd. (the original offer was made by Doulton).

Why have small firms not survived in the sanitary ware sector of the industry? (Takeovers are not an explanation as they are a means rather than a cause.)

We may take the case of Doulton first. Why did they acquire Johnson & Slater Ltd.? They considered that their Stoke-on-Trent factory did not provide sufficient output to offer their customers a full range of products. The capacity of their existing factory was limited because land was not available for expansion. A factory in Kent was regarded with particular favour because it would lead to a saving in transport costs; they would be higher if the national market were served from only one centre. Other factors too, presumably, played a part. Doulton, with their long tradition of pioneering sanitary appliances, were confident that their managers were as good, if not better than, other managers in the industry (nearly all pottery managers seem to think this way). Doulton must benefit, however, from having the research division of the Group to draw on as well as other management services, in particular, management training and recruitment, financial support and control, and public relations. To build a new or an additional factory would, in Doulton's opinion, have cost too much, so the takeover was the preferred alternative. In the event Doulton found that bringing the facilities of the Kent factory up to their desired standards was a severe strain on their management and financial resources. (Looked at the other way round—the takeover enabled surplus management capacity to be fully utilised.) An important consideration was to achieve longer runs of particular items in each factory to reduce unit costs. Without any control of prices it was felt that an increase in size would help to restrain costs and sustain profits. That an increase in size brings benefits was explicitly stated by the Chairman of Armitage Shanks Ltd. 'During 1970 our group of companies, in common with the general trend, have been faced with substantially increased costs, especially wages and salaries, which have not been entirely offset by increased selling prices or higher productivity. In the long term the Group will greatly benefit through its ability to offer a wide range of its own manufactured products for the building industry.'

The firm of Edward Johns & Co. Ltd. was registered in 1907. It moved from Stoke-on-Trent to Armitage, to avoid a labour dispute it is said, and in 1960 the name was changed to Armitage Ware Ltd. Its two recently built factories and office block and warehouses, situated in the country, suggest a successful company, and its trading and profit figures confirm the impression. The merger with Shanks provides Armitage Shanks Ltd. with a dominant position in the industry. The two firms in the takeover continue to trade fairly independently. The issuing of a joint catalogue and the amalgamation of their London showrooms were short-term benefits; other benefits may be visualised, however long they take to mature. Shanks Holdings Ltd. developed the battery method of production which is now being introduced into all of Britain's sanitary factories. Like tunnel kilns replacing bottle ovens, the old method must give way to the new and superior one. The battery was devised by a marine engineer while he was spending time ashore working in Glasgow on sanitary ware. Instead of each caster filling individual moulds, batteries of moulds are connected to pipes and slip circulates continuously for three or four hours until the moulds are properly filled. The battery is usually switched-on automatically in the early morning and when the casters arrive the moulds are ready to be emptied. Obviously the production process is substantially speeded-up by the innovation. At present the battery system supplements rather than replaces traditional methods.

Ideal–Standard Ltd. was founded in 1894 as the National Radiator Co. and in 1934 the name was changed to Ideal Boilers and Radiators Ltd. The Company is controlled by American Standard Inc. and it may be assumed that the Company was not restricted in its growth by a shortage of finance. In 1970 two main divisions of the company were formed: 'One is concerned with heating and ventilating and air conditioning products and the other comprises vitreous china, porcelain enamelled cast iron, plastics and brass sanitary fixtures and fittings.' The firm's expansion in ceramic sanitary ware production is to be understood in terms of the overall expansion in a number of related fields. In 1968 the firm had 3700 employees and turnover was £12 million, 46% of which was accounted for by sanitary products, in the same year the latter accounted for 39% of the Group's profits. The figures suggest that profits on ceramic ware were modest. The profits on all sanitary products included those on plastic and brass fittings. The Group as a whole has enjoyed a good profit record.

In 1896 a tank measuring 500 gallons capacity made in Mr. Twyford's factory 'created a great sensation' at a laundry exhibition. In the 1890s Thomas William Twyford had introduced the manufacture of large pieces of sanitary ware to Stoke after having examined production techniques in Scotland: 'The crux was the size of the pieces, and Mr. Twyford had in his mind such an innovation in this direction as staggered his new hands no less than it was the despair of his old ones, and during the period of experiment taxed his own patience and perseverance, though he never doubted of ultimate success. . . . If Mr. Twyford was at the outset far from conscious of all the difficulties and troubles that were in store for him and the enormous monetary losses he would have to suffer in starting his new enterprise, he never once exhibited a sign of doubt or retreat.'* A pottery had been established in Stoke by Joseph Twyford before 1790.

After 1945 Twyford's two factories were extensively modernised and tunnel kilns were installed. A complicated system of overhead transporters was later devised to link the workrooms at various levels and reduce the size of the labour force engaged in transporting raw materials and ware from process to process. The sales force established particularly good relations with merchants and export markets were not neglected. In recent years the firm has offered to supply all items

---

* Joseph Hatton, *Twyfords: a Chapter in the History of Pottery*, J. S. Virtue & Co. Ltd.

Plate 9.5   Kiln room at Twyfords (Holdings) Ltd.

in their catalogues on two weeks delivery. It has always been the
firm's policy to manufacture a particularly wide range of items. As
the market for sanitary appliances weakened, Twyfords were particularly successful in sustaining their sales and their profits. This was
probably a consequence of Twyford's close attention to customer
needs and their willingness to supply short-runs of non-standard items.
Also, unlike rival firms, managers were not diverted from attending
to customer needs by acquisitions of other companies. A third factory
was erected by Twyfords in the country north of Stoke-on-Trent
and a warehouse there is used by all the production units. An offer
made for the Company in 1971 by Glynwed Ltd. was not accepted.
Subsequently, Twyfords Ltd. were acquired by Reed International
Ltd. The principal activities of the Reed Group are: building products,
decorative products, paper and paper products, and printing and
publishing (including the publishers of this book). In their 1972
Annual Report the Chairman of Reed International stated: 'Twyfords
Holdings Ltd. was acquired in 1971, bringing into Reed International
one of the most highly reputed manufacturers of ceramic building
products. We were thus enabled to fulfil our plan of establishing
a new Main Division 'Reed Building Products'. In addition to Twyfords, Reed International also own Polycell which, among other things,

manufactures ceramic tiles. However, tiles and sanitary ware operate in two fairly self-contained divisions of the Group.

Table 9.2 SANITARY WARE MANUFACTURERS: PROFITS 1965–1971

Year	Capital employed (£'000)	Pre-tax profits (£'000)	Profits as % of capital employed (%)	Sales (£'000)
	Armitage–Shanks Group Ltd.*			
1965	2 398	544	22·7	—
1966	2 344	426	18·2	—
1967	2 404	336	14·0	—
1968	2 894	562	19·4	4 329
1969	2 843	643	22·6	4 822
1970	8 078	803	10·0	9 990
1971	8 401	1 304	15·5	12 320
	Twyfords Holdings Ltd.			
1965	2 986	626	21·0	3 446
1966	2 890	445	15·4	3 256
1967	2 979	461	15·5	3 378
1968	3 168	583	18·4	3 788
1969	3 490	649	18·6	4 455
1970	3 820	709	18·6	5 031
1971	4 126	957	23·2	—

Source: Moodies' Services Ltd.
*Armitage Ware only until 1970.

FIRECLAY

In 1963 the *Census of Production* listed 14 establishments engaged in making sanitary fixtures from fireclay; before World War II there were many more. At present all but three of the sanitary fireclay works are controlled by manufacturers of vitrified china. They use fireclay to manufacture large pieces which, at present, cannot be made of vitreous china or could only be made at extremely high cost. Fireclay is used for mortuary slabs, urinals, shower trays, specialised hospital fittings and industrial sanitary wares. Gradually fireclay has been replaced by stainless steel or constructions using pieces of vitreous china. Users have found that lighter pieces will do where before more substantial pieces were deemed to be essential. Firms continue to make fireclay pieces as it completes their product range. Understandably firms like to manufacture glazed fireclay products which, sledge-hammers apart, are liable to last for hundreds of years without developing signs of ware. Fireclay pieces are frequently made-to-order and the final products are completed by means of saws and planing machines. Metal and plastic fittings are frequently added. Chipping, boring, filing and sawing are unusual in the pottery industry; they also occur in the last stages of the manufacture of large electrical insulators.

ELECTRICAL WARE

MANUFACTURE AND TESTING

Large, and small, insulators are made from a slip composed of feldspar and quartz sand combined with ball clays and china clay. For a large insulator each section is made in a plaster mould which determines the outside shape. The centre part is cut out and 'while the clay mould revolves the shed section is formed by a metal tool manipulated up and down by the jolley operator. When an insulator is tapered each shed may require a different mould'. Making large pieces of clay to accurate specifications requires a high degree of skill. At the forming stage the clay contains 21%–22% of water; when partially dried to 16%–17% of water the various parts shrink away from the moulds. Some insulators are formed in one piece by turning on a lathe. The shape is controlled by a template which is followed by the operator. 'More complex sheds are built up one by one on rotating platforms. The faces to be joined are carefully trued and then joined with porcelain slip. There is a joint for every shed and every joint must be perfect.' After being fettled the insulators are moved into special dryers and, after drying, are dipped into a tank of glaze slip. 'This slip is a suspension of finely milled feldspar, calcium carbonate and an oxide stain to produce the required glaze colour. Partly fired granules of a special porcelain are attached to the plain ends by an adhesive-glaze mixture. These provide a grip for the cement in later porcelain-to-metal or porcelain-to-porcelain assembly operations.' Glaze may also be applied by means of spray guns. Firing to about 1200 °C and cooling take about six days. 'During firing the shrinkage is about 16% in height and 10% in diameter. After firing has been completed, some grinding is necessary to achieve the correct length and to make the end faces flat, parallel to each other at right angles to the longitudinal axis. Metal flanges are now attached to the ends of the insulator using a high-strength Portland cement and sand mixture. After the cement is cured and the insulator has been electrically, mechanically or hydraulic pressure tested, it is ready for despatch and assembly in customer's equipment'.

The testing of large insulators is both an expensive and spectacular operation. They must be perfect and they must be capable of standing-up to adverse weather conditions, including dust storms and ice blizzards. In addition they must resist corrosive material in the atmosphere.

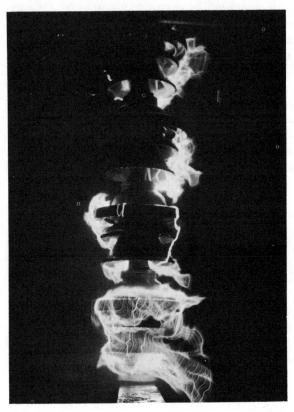

Plate 9.6   Insulator under test (Doulton Industrial Porcelain Ltd.)

The *Census of Production* in 1963 listed 18 firms which manufactured electrical ware; the value of their sales was £9 872 000. Two firms were responsible for at least two-thirds of total sales: Allied Insulators Ltd. (the name adopted by Bullers Ltd. and Taylor Tunnicliffe & Co. Ltd. when they amalgamated in 1959) and Doulton Insulators Ltd. Allied Insulators and Doulton between them manufacture most of the high tension insulators produced in Britain. Allied Insulators also produce large quantities of low tension insulators, the remainder being produced by the smaller firms in this sector of the industry. One firm, George Wade & Son Ltd., combines the production of insulators and ornamental and hotel ware.

From 1945 until 1966 there was a sustained boom on the home market for electrical insulators. It was caused by extra capacity being installed to generate and distribute electricity. In Britain, between 1953 and 1966 the growth of demand for electricity was just over 9·4% per annum. An additional boost to demand for insulators arose from the decision to construct the super-grid in Britain which now operates at voltages of 400 kV and 275 kV. After 1966, when the growth of demand for electricity fell and the super-grid and the electrification of the railways from London to Crewe had been completed, the demand for insulators fell drastically. The replacement demand for insulators is a modest one; they are manufactured to high standards in terms of quality and deterioration is an extremely slow process.

The severe fall in demand coincided with an increase in capacity by the firms in the industry. An increase in orders had been expected as a consequence of the Government's plans to achieve a growth rate of 4% per annum after 1966. The orders were expected because the Department of Economic Affairs and the Ministry of Technology had appealed to the manufacturers to increase capacity to cater for their 'planned' growth in electricity consumption. Businessmen can be unduly critical and suspicious of government actions; however, the manufacturers of ceramic insulators have genuine grounds for complaint. The lack of orders for insulators was only one small aspect of a lack of orders for heavy electrical equipment. One consequence of excess capacity in the sector of the engineering industry which makes heavy electrical equipment was the takeover of the English Electric Company by the General Electric Company and the rationalisation of production units that followed. Rationalisation did not occur in insulator production, at least not before 1972; instead, this sector of the pottery industry has had large areas of factory space lying idle. Manufacturers of electrical equipment have always sub-contracted to the pottery industry for large insulators. One manufacturer did attempt to make them rather than buy them when supplies were particularly scarce, but the venture was not a success. Presumably it was not possible to achieve the high quality standards which are essential in a short period of time.

Table 9.3 shows the value of sales and the value of exports of electrical ware; also the value of capital expenditure on fixed assets by the Electricity Council and the Electricity Boards each year from 1961

Table 9.3 SALES AND EXPORTS OF ELECTRICAL WARE (1951–1970) AND CAPITAL
FORMATION IN ELECTRICITY GENERATION AND DISTRIBUTION (1961–1970)

Year	Electrical ware		Electricity supply Capital expenditure on fixed assets ($£$ million)
	Sales ($£$'000)	Exports ($£$'000)	
1951		1 146	—
1952		1 248	—
1953		936	—
1954		968	—
1955		1 076	—
1956		1 343	—
1957		1 167	—
1958		1 300	—
1959		1 187	—
1960		1 420	—
1961	8 297	1 569	322
1962	8 707	1 399	373
1963	9 108	1 158	469
1964	10 673	1 505	562
1965	11 124	1 610	591
1966	11 854	1 519	661
1967	10 505	893	599
1968	9 510	1 292	469
1969	8 856	1 452	392
1970	10 196	1 906	375

Sources: British Ceramic Manufacturers' Association, and *Handbook of Electricity Supply Statistics* (1971 Edition).

to 1970. It may be seen from the table that sales, in value terms, declined after 1966 in spite of improvements in export sales. The severe decline in capital formation in the electricity industry after 1966 explains why the home market has been so weak. In March 1971 the Chairman of Allied Insulators Ltd. reported that 'the section of the Company which caters for generation and distribution of electricity for the Central Electricity Generating Board is now almost without orders for the home market and is depending in 1971 to an even greater extent upon direct and export orders which carry only marginal profits'. The Chairman of Doulton & Co. Ltd. also reported an unsatisfactory situation in insulators: '1970 proved as difficult a year for our insulator factory at Tamworth as we had expected. In the first six months a loss was made. However, a more careful selection of export orders taken, together with the introduction of measures to reduce overheads and contain cost initiation, led to an improvement in the second half of the year. This was a somewhat better performance than we had budgeted for, but we consider conditions in 1971 will remain difficult'.

Table 9.4 shows the extent to which the profit performance of Allied Insulators Ltd. was adversely affected by the fall in home demand (profit figures from Doulton are not available for insulators only). The two companies, Bullers and Taylor Tunnicliffe, have operated fairly independently since they amalgamated in 1959. Bullers operate two factories in Stoke-on-Trent and one at Tipton in the Black Country, and Taylor Tunnicliffe operate a large factory at Stone, some 12 miles south of Stoke-on-Trent, and three additional factories in Stoke-on-Trent. Over the period 1967–1970, Allied Insulators accounted for 53% of total sales by the industry. In 1973 Allied Insulators Ltd. acquired the Ceramic division of Plessey & Co. Ltd. This enabled the firm to extend its product range.

Table 9.4 ALLIED INSULATORS LTD: PROFITS 1961–1970

Year	Capital employed (£'000)	Pre-tax profits (£'000)	Profits as % of capital employed	Sales (£'000)
1961	3 143	658	16·8	—
1962	3 325	722	21·7	—
1963	3 433	740	21·6	—
1964	3 712	1 035	27·9	—
1965	4 097	1 087	26·5	—
1966	4 392	1 273	29·0	—
1967	4 183	762	18·2	5 378
1968	4 110	331	8·3	5 020
1969	4 131	220	5·3	4 937
1970	4 332	431	9·9	5 433

Source: Moodies' Services Ltd.

The manufacture of high tension insulators requires plenty of space for making, drying, glazing, firing and testing. Highly specialised equipment is not used to make insulators, so it should be possible to switch production from insulators to sanitary ware and other industrial ceramics. A switch to sanitary ware did not occur, presumably because the demand for sanitary ware slackened at the same time as the home demand for insulators slumped. Also a change-over, except for Doulton, would have been difficult, as the channels of distribution for the two groups of products are entirely different. The firms probably calculated that by the time a change had been made the demand for insulators would have recovered. The firms know that demand will recover some day; the problem is how long they can afford to wait.

INDUSTRIAL CERAMICS

Industrial ceramics are used as materials and equipment in many manufacturing industries. The products cover a wide range, for as new products are added only a few are removed. They include nose cones for space rockets, moulds for the blades of jet engines and turbines, and balls and chips for use as polishing media. Porous ceramics are widely used as filtering media, while ceramic materials are used for containers to hold liquids and solids which would attack metals and other materials. Components may be made of ceramic materials where industrial processes demand abrupt changes of temperature, as in the hardening of metals, or where hot substances are handled or moulded. The two leading producers are Doulton Industrial Products and Royal Worcester Ltd. Allied Insulators Ltd. is one of a number of firms which manufacture some industrial ceramics in conjunction with the products in which they specialise. It has been the policy of Wade Potteries Ltd. since the 1950s to move out of domestic ware into industrial ceramics.

The output of industrial ceramics, excluding electrical and sanitary ware, was only a few million pounds sterling in 1971. However, output may be expected to expand considerably once growth is restored in the economy as a whole. An estimate by the Battelle Memorial Institute is that the world market for industrial ceramics at the end of the 1960s was increasing at 15% per annum.

DOULTON INDUSTRIAL PRODUCTS

It is a task for scientists in specialised laboratories to develop new industrial ceramics. Until recently, lavish facilities for research and development were provided by Doulton at Chertsey in Surrey, where materials and processes were developed for domestic and sanitary ware as well as for industrial ceramics. The real challenge to the engineers, chemists and physicists at Chertsey, however, was provided by the need to develop new types of industrial products. Progress was made in the use of silicon nitride, which has the admirable attribute of being virtually impervious to changes in temperature. Capable scientists and technologists were attracted to Chertsey by the challenges offered by industrial ceramics. Once there, they were available to work on more mundane research tasks which arose in the electrical, sanitary and domestic ware sectors of the Company. For years the sanitary and domestic ware divisions of Doulton enjoyed the benefit of having an experienced research team available to help with their production problems. Unfortunately, with the advent of severe stagnation in the British economy after 1968, the markets for

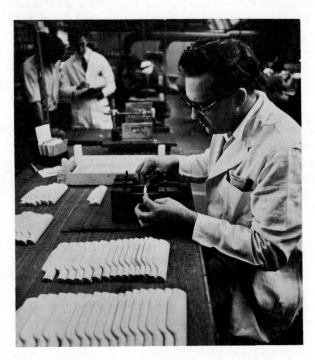

Plate 9.7   A gauging operation being carried out: Quality control has to be rigidly exercised throughout the manufacture of Doulton preformed ceramic cores (Doulton Industrial Products Ltd.)

new industrial products failed to materialise and, in consequence, work at Chertsey was severely curtailed. Without a good profit being earned by industrial ceramics the other divisions of the Company could not be expected to carry the substantial research and development costs. The curtailment of research and development in the pottery industry in 1971 reflected similar cut-backs in most other industries. In that year large numbers of technologists and scientists were rendered redundant.

Industrial ceramics were manufactured by Doulton at their electrical porcelain factory at Tamworth until 1968, when the site and buildings of a tile factory were purchased near Stone, south of Stoke-on-Trent, and Doulton Industrial Products commenced operating there. Subsequently, after Doulton had been acquired by S. Pearson & Sons, it was decided to operate industrial ceramics, sanitary ware and electrical porcelain as one unit within the Pearson Group, and Doulton Fine China Ltd. and the various Allied English companies as another fairly independent unit. It would be fair to say that once Doulton developed the tableware and ornamental side of the Company it never had much in common with its other activities.

ROYAL WORCESTER LTD.

The Royal Worcester Porcelain Co. Ltd. manufactures some laboratory and industrial equipment in its Worcester tableware factory, using hard porcelain bodies. However, the Group operates an industrial ceramics factory at Tonyrefail in South Wales. While Doulton specialises in silicon nitride, Royal Worcester has developed alumina as well as steatite and oxide ceramics. Their main customers are textiles, aircraft, chemical and engineering companies. Gas igniter leads for use with natural gas have been developed. The company became involved with electronics in wartime, and in 1946 an interest in Welwyn Electric Ltd. was acquired. In 1953 the Company was taken over completely. Now electrical engineering is a division of the Group which operates independently from its pottery division.

WADE POTTERIES LTD.

For over 100 years members of the Wade family have made pots in Burslem, Stoke-on-Trent. This independent group of companies has progressed by developing new products and techniques and abandoning its traditional products.

The factory space in which A. J. Wade Ltd. manufactured decorative glazed tiles has been converted to produce higher temperature products such as porcelain and alumina. A variety of items are die pressed, including top quality substrates in alumina for the production of micro-electronic circuits. Wade Heath & Co. Ltd. has ceased to produce conventional earthenware tableware and now specialises in advertising earthenware—jugs, ash-trays, mugs and bottles, etc. In addition, there is a factory in Portadown in Northern Ireland where ornamental porcelain and electrical porcelain, etc. is manufactured. It is their wide product range which enables Wade's to claim that they 'specialise in diversity'.

In 1971 the Group's turnover was £2 590 000. It was split, approximately, 50% industrial ceramics, 25% advertising products and 25% ornamental porcelain. The profit rates on capital employed, between 1967 and 1971, were 9·5%, 18·4%, 14·3%, 23·3% and 30·8%. The profit rates indicate that Wades are one of the few remaining independent and successful businesses in the pottery industry.

Industrial ceramics are manufactured from a variety of bodies and firing temperatures vary considerably. This renders production and warehouse control a complicated matter. On the other hand, new products can be added to the existing product range at low cost and this contributes towards effective marketing. Once a die has been made of steel or tungsten, thousands of identical items can be made on hydraulic or mechanical presses. Production is not automatic, as it is in the production of tiles. Care must be taken in releasing items from the dies—unfired pressed dust is fragile. In contrast with a steel die, a plaster mould suffers severe deterioration after only a dozen or so items have been made. In addition, a die can impose much sharper detail on an item than is possible when a mould is used for casting. Virtually anything can be made of porcelain using the die-pressing technique, provided the item is small, and Wade's take full advantage of this fact. Often, however, to fulfil a customer's requirements a lot of ingenuity is required to discover the right blend of ingredients and temperatures. Also, customers are forever redesigning their products, so that the component manufacturer may find dies and production processes redundant. Long production runs are essential and dies are extremely expensive. What may appear to be a simple production operation, in practice is extremely exacting. Without good management profits would quickly disappear. The production of die-pressed items is supplemented by some items which are cast. The dies are made by highly-skilled fitters and the factory has many of the characteristics of an engineering workshop. In addition to industrial ceramics, premium give-aways are made on the presses in hundreds of thousands, week in and week out. There are small animals and birds and Disney and fairy-tale characters which are hand-painted. They are inserted into packets of tea and coffee and breakfast cereals so that children will insist of their parents purchasing more of the same. By means of mass-production techniques, perfect porcelain figurines are sold in bulk for only one or two pence each. In recent years, Wade's have tended to specialise in products which are not produced by other pottery manufacturers; however the firm faces substantial competition from firms which make similar products in non-ceramic materials, mainly plastics. By specialising in top quality and therefore relatively expensive products the firm has tended to avoid the excessive competition which dominates the production of cheap electrical components. Specialised refractories for use in gas fires and appliances are also produced at Burslem. In 1964 William Kent (Porcelains) Ltd. were purchased and the works were closed after equipment had been transferred to Wade's.

Wade Heath & Co. Ltd. was one of an excessive number of small earthenware firms which continued in production into the 1950s. It was recognised, however, that there was little to be gained by offering similar ware to that offered by the larger earthenware manufacturers who included Johnson Brothers and Ridgway. It was decided to substitute advertising earthenware products for the conventional tableware products. They now enjoy an established position in providing ware for distillers, brewers, tobacco manufacturers and airlines. A subsidiary company was launched to do the marketing and Wade (PDM) Ltd. offers for sale items made of glass, plastics, rubber or metal which are manufactured outside the Group as well as the products of the parent company. One further field of activity for the Wade Group is decorative transfers. They are used mainly for their own products and are supplied to outside customers too.

A branch factory was located in Northern Ireland at Portadown shortly after the end of World War II. The past chairman of the Company, Col. Sir G. A. Wade agreed to go to Ulster after Sir Stafford Cripps had insisted that to do so would be in the national interest. It was the declared aim of the 1945 Labour Government to 'attract' industry to regions where unemployment was high. The original output was mainly small insulators for electric cookers and other appliances. Subsequently a wide range of high and low tension insulators was developed at Portadown, mainly for switchgear and for use by the Post Office. Ornamental ware is also produced at Portadown. As becomes Ulster's only pottery, the ware is mostly mottled green and decorated with traditional Irish symbols and designs. The ware sells well in Ireland, mainly to tourists, and a high proportion is exported to the United States and Canada. There is no obvious advantage obtained by splitting the output of a medium-sized firm between England and Northern Ireland. However, the introduction of new skills to the Province has brought substantial benefits to Northern Ireland.

Alumina, which is supplied under an exclusive licencing agreement with the Western Gold & Platinum Company of California, is processed both at Portadown and at Burslem. The policy is to concentrate on the top end of the market producing alumina components of the highest purity intended for the most exacting applications. The Burslem substrates for use in the micro-electronics industry provide 'permanent dimensional stability, high thermal conductivity, chemical inertness and exceptional electrical properties at elevated temperatures'. Other products made of alumina are fuel element supports for nuclear reactors, cylinder liners, and parts which must be resistant to abrasion. Alumina products are expensive and are purchased only when cheaper substitutes would not be entirely satisfactory. Pieces made from alumina have twice the strength of conventional porcelain pieces and their hardness is comparable with that of sapphire. The marketing of Wade's alumina products is undertaken by the English Glass Company Ltd. of Leicester. Royal Worcester too makes alumina products, however their products do not compete directly with those produced by Wade.

Wade is an example of a medium-sized firm which has progressed by paying careful attention to marketing and by blending engineering techniques with traditional pottery skills. Whilst we are inclined to point to Wade's active management team to explain their success, they, in contrast, consider that their success has been firmly based on the resilience and loyalty of their labour forces in both Burslem and Portadown.

Part IV

LABOUR

10

WORKERS,
WAGES AND PRODUCTIVITY

A theme which recurs in this book is that as wages increased in the pottery industry, so manufacturers substituted capital for labour and improved the efficiency of their production and marketing operations. This form of presentation places undue emphasis on management and capital as the agents of change and suggests that the workers are merely passive participants. In practice the industry depends for its survival on the skill, goodwill and perseverance of its labour force. Here we attempt to present the pottery workers, their achievements and problems and their union. In July 1970, the National Board for Prices and Incomes reported to Parliament on 'Pay and other terms and conditions of Employment of Workers in the Pottery Industry'; we have drawn extensively in this chapter on this valuable source of information.

EMPLOYMENT

Figure 10.1 shows employment in the pottery industry in Great Britain, month by month, for men and women for the years 1948–1968. It may be seen that throughout the period women outnumbered men and that male employment contracted less than that of women over the period. The month for peak employment was September 1951,

204

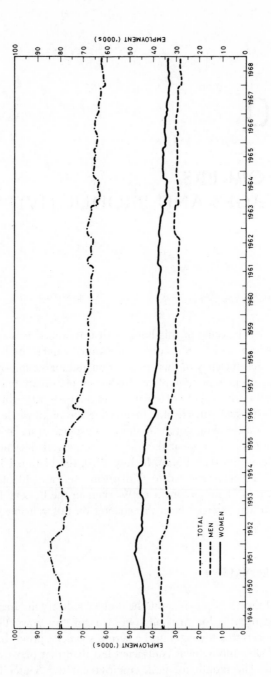

Figure 10.1 Employment in the pottery industry 1948–1968

when 85 300 persons were employed; by September 1968, employment had fallen to 62 000 persons; and by September 1971, there had been a further fall to 57 000 persons, composed of 28 100 men and 28 900 women. The fall in the labour force occurred in spite of a fairly steady rise in real output and it was mainly a consequence of mechanisation and rationalisation. It is unlikely that output will increase substantially in the foreseeable future (there do not appear to be rich untapped markets at home or abroad) and mechanisation and rationalisation will continue. One must therefore expect a further fairly substantial contraction of the labour force in the pottery industry over the next 20 years.

THE OCCUPATIONS OF MEN AND WOMEN

Table 10.1 shows the labour force classified by occupation group for pottery establishments in Great Britain in January 1970. (The response rate by establishments to the survey on which the table is based was 78·3% so the figures are intended to illustrate only relative magnitudes.) Examination of the figures reveals that women workers are concentrated in making domestic ware, electrical porcelain and tiles (their jobs there frequently entail minding machines), in warehouses (packing), in fettling domestic ware (smoothing down biscuit ware) and in decorating. In contrast men dominate clay preparation, mould making and operating the kilns. Men also outnumber women in electrical porcelain and sanitary ware production.

SKILL

We associate skill with the potter turning a pot on the wheel. A few men still operate wheels in large factories in the pottery industry today and they serve as a reminder that mechanisation is not yet complete in the industry. They preserve an image and they interest visitors. They can also pay their way by specialising in making a few prestige lines for their firms. Pots can be made just as well by machines (cups, saucers, plates and bowls); in fact, once the machines are operating properly ware can be made with much greater consistency with their aid than on the wheel. Once the objective is mass-production of identical pots then machines are supreme. If the customer wants one-off ware then craft potters are available, usually in small firms, to satisfy the need. In many occupations in the pottery industry the standardisation of

Table 10.1 THE LABOUR FORCE OF THE POTTERY INDUSTRY ('000s) 1970*

Occupation group†	Domestic ware Men	Women	Electrical porcelain Men	Women	Tiles Men	Women	Sanitary ware Men	Women	Ancillary products Men	Women	Total Men	Women
Adults												
Clay prep. workers	0·7		0·2		0·2		0·2				1·3	
Mould makers	0·7	2·5		0·8		0·6	0·1				0·8	3·9
Makers	1·9		0·3		0·2	0·1					2·4	0·1
Kiln workers	1·8	4·5	0·2	0·1	0·8	0·9	0·3				3·1	5·6
Warehouse workers	1·2	2·3		0·1	0·4		0·3	0·1			1·9	2·4
Fettlers		0·8	0·1			0·1					0·1	0·9
Glazers	0·5	5·3	0·1		0·2		0·3				1·1	5·3
Decorators	0·4						0·1		0·2		0·5	
Labourers	1·2		0·1		0·3		0·4		0·2		2·2	
Ancillary workers‡	0·6	0·3	0·3		0·7		0·5	0·1	0·2		2·3	0·4
Other	0·1						0·1		0·2		0·3	
Aged 15–20	1·4	1·0		0·2	0·5	0·4	0·2				2·1	1·6
Total workers aged 15 years and Over§	10·5	16·7	1·3	1·2	3·3	2·1	2·4	0·2	0·6		18·1	20·2

Source: *Pay and Other Terms and Conditions of Employment of Workers in the Pottery Industry*, National Board for Prices and Incomes: H.M.S.O., Report No. 149, Cmnd. 4411, Table 3, London (July 1970).

*The table does not include stoneware and industrial ceramics manufacturers or tile slabbers.
†Where no workers are shown in a category it is possible that workers in that category have been counted elsewhere. This arises because the table is based on data for men and women separately and not all the occupation groups appeared in both sets of data.
‡Canteen workers, cleaners, drivers and maintenance workers (building craftsmen, electricians, fitters, etc.).
§Figures contained in the National Board for Prices and Incomes Survey. The total labour force reported by the Department of Employment was 28 700 men and 31 000 women (July 1970).

materials and the use of machinery has meant that skilled labour has been replaced by semi-skilled labour. In decorating, for example, coloured patterns are applied by means of lithos, and imprints can be taken off engraved copper plates and applied to ware by means of parabola-shaped gelatin pads. More and more hand-painting is restricted to a few fine china patterns. To operate a machine tends to require little skill; however, the workers who set and maintain the machines require engineering skills. Highly-skilled pottery workers, of course, remain; they make moulds, undertake difficult moulding operations, apply glazes, operate the kilns, load and unload ware into and out of the kilns, inspect ware for faults, work as gilders with silver, gold and platinum, and operate warehouses and packing operations. More and more workers, however, carry out detailed instructions and use their judgment less and less because bodies, glazes and colours are strictly controlled. The skill which consisted largely in making adjustments to offset variations in raw materials, etc. is now, because of quality control, no longer required for many operations.

Pottery workers would be inclined to agree that skill is not what it was; however, they would argue strongly that a high degree of skill remains and that they have an intuitive feeling for clays, glazes, shapes and colours and quality which is of immense benefit to management. This feeling and respect for clay is not shared by other workers and must be gained by experience over the years. They would argue further that only skilled workers could operate with the consistency and speed required by the management and the machines, and for many seemingly semi-skilled operations a high degree of manipulative skill is required as in gilding and decorating. This is a different meaning of skill than the one we have used, and we would agree that, in their way of viewing the situation, a high degree of skill is still required of pottery workers.

It is not possible to measure skill adequately by means of wage rates. On this point the Prices and Incomes Board commented as follows: 'Over the post-war period, there has been a wave of new processes together with the mechanisation of existing ones, resulting in substantial changes in the jobs performed. The effect has been to produce an earnings structure which in many respects is not in accord with the occupational groups in the agreement'. (The agreement relates to the comprehensive wage structure for the industry which sets minimum rates of pay and conditions of employment.) Wage rates reflect supply

---

*National Board for Prices and Incomes, Report No. 149, H.M.S.O., London, para. 47 (July 1970).

and demand conditions, conventions, bargaining strengths, degrees of unpleasantness, responsibility and monotony as well as skill. Average earnings too are an inadequate measure of skill as they reflect, among other things, hours worked and intensity of work, which may be reflected in piece-work earnings. *Table 10.2* shows average earnings in January 1970 for selected occupations. It is reasonable to assume that the highest average earnings reflect to some extent a higher skill

Table 10.2 OCCUPATIONS RANKED BY AVERAGE HOURLY EARNINGS (NEW PENCE): ADULT MEN AND WOMEN (JANUARY 1970)*

Women		*Men*	
Average earnings per hour	*Occupations*	*Average earnings per hour*	*Occupations*
0·26	Potters' assistants	0·42	Labourers and glaze mill workers
0·30	Overlookers and dipping house workers	0·43	Carriers
0·31	Glost and biscuit warehouse workers	0·45	Warehouse workers
0·33	Decorators and transferers	0·46	Mill labourers
0·34	Silk screen decorators	0·48	Slip makers
0·35	Fettlers	0·49	Kiln assistants and slip-house workers
0·36	Gilders, liners and electrical porcelain makers	0·50	Maintenance fitters and printers
0·37	Spongers, tile press attendants and towers	0·51	Cup and bowl makers
0·38	Handlers	0·52	Dippers, platemakers, enamel placers and drawers
0·39	Casters	0·53	Placers (continuous ovens) and kiln firemen
0·41	Flower makers	0·55	Polishers
		0·56	Placers (intermittent ovens), pressmen and casters
		0·58	Casters
		0·59	Placers (intermittent trucks)
		0·60	Packers
		0·62	Mould makers
		0·65	Decorators
		0·79	Sanitary glaze sprayers
		0·81	Sanitary casters

Source: National Board for Prices and Incomes, Report No. 149, H.M.S.O., Table 10, London (July 1970).
*Average earnings per hour can be high for relatively unskilled work when long hours are worked.

content of occupations. It may be noted that 'maintenance fitters' in a non-pottery industry skilled trade earned less per hour than placers, mould makers, casters, etc. There are occupations in the pottery industry which are obviously highly-skilled; flower makers, sanitary casters, mould makers are examples. However, which occupations should be added to this list and where the line should be drawn between skilled and semi-skilled must remain a matter of opinion for the present. All we can say with certainty about skill is that in recent years in the pottery industry it has been diluted; we cannot say by how much and how fast. One personnel manager pointed out to us that this trend is likely to be reversed in the long-run. He expects there to be a substantial reduction in the semi-skilled labour force accompanied by the recruitment of technologists, computer programmers, chemists, engineers and even sociologists. The consequence of new technology is likely to be a smaller yet a more skilled labour force.

LOW PAY?

There is a widely held belief that the pottery industry in Stoke-on-Trent pays particularly low wages and that this explains why pottery is a labour intensive industry. The belief is completely untrue. In a full employment economy (due allowance being made for 'temporary difficulties') an industry which consistently paid particularly low wages would lose its labour force. An industry, particularly one situated mainly in the West Midlands, must, since 1945, have paid wages that are in the line with other industries and occupations. Admittedly the pottery industry has reduced the size of its labour force in recent years; this was a consequence of a fall in the number of jobs—it was not an effect of low pay. Pottery is a labour intensive industry, not because the wages of its workers are unduly low, but because it is possible to obtain high prices for ware with a high labour content. There is some evidence that pottery wages were relatively low before 1939. B. R. Williams has pointed out that the average weekly earnings of male and female pottery workers increased to a considerably greater extent between 1938 and 1954 than comparable earnings of all male and female workers.*

In 1964 the Department of Employment calculated the average total labour cost per employee for various industries.† Labour cost

*'The Pottery Industry', Volume 11 *The Structure of British Industry*, a Symposium edited by Duncan Burn, Cambridge University Press, p. 297 (1958).

†Department of Employment and Productivity, *Labour Costs in Great Britain in 1964*, H.M.S.O., London.

included wages and salaries, national insurance contributions, private social welfare payments and payments in kind. For all manufacturing industries in 1964, labour cost per employee was £886; in contrast, in the same year, labour cost per employee in the pottery industry was only £622. (What is labour cost to the employer is, of course, income for the employee.) Only clothing and footwear had a lower labour cost than the pottery industry (£561). Even textiles and leather and fur were higher (£697 and £745). Industries that were much higher included engineering and electrical goods (£902), paper and printing (£956), metal manufacture (£947), vehicles (£1056) and chemicals (£1089). Low labour cost per employee is not the same thing as low pay and perhaps the two measures tend to be confused. The relative low average labour costs per employee in the pottery industry reflect, among other things, the high proportion of the work force that are women, the high proportion of semi-skilled and unskilled jobs, the high incidence of part-time working and the extent to which full-time workers work relatively short hours. We must compare women's earnings in pottery with their earnings in other industries, and similarly for men, to obtain more satisfactory comparisons.

In their publication, *General Problems of Low Pay*,* the National Board for Prices and Incomes provided a table of comparative rankings by industry (lowest earnings first) by average weekly earnings of men manual workers. Pottery was ranked 37 in 1960, it rose to 59 in 1967 and it was 54 in both 1968 and 1969. In 1967 National Government Service was 1 and printing, publishing and newspapers was bottom, i.e. well paid (128). In the same year industries with lower earnings than pottery included: leather goods, brushes and brooms, electricity, timber, woollen and worsted, bedding, etc., many sectors of the clothing industry, bread and flour, confectionery, paint and printing ink, footwear, and explosives and fireworks. For men, the pottery industry ranks higher than many industries which have never been considered to be low pay industries. In terms of pay the pottery industry would appear to be just slightly below average. (If we look at the dispersion about the average we find that, compared with other industries, the pottery industry has relatively few workers who receive particularly low pay and relatively few workers who receive particularly high pay. It is the latter factor which contributes to keeping-down the average.) Other evidence suggests that women pottery workers are in a slightly more elevated position.

The Department of Employment carried out a sample of the earnings

*National Board for Prices and Incomes, Report No. 169, Cmnd. 4648, H.M.S.O., London (April 1971).

of employees in employment in Great Britain in April, 1970.* For the pottery industry the average (median) gross hourly earnings of full-time manual men (aged 21 and over) was 55·5 new pence which is fairly close to the 59·0 new pence for all manufacturing industries. For full-time manual women (aged 18 and over) the average gross hourly earnings were 34·5 new pence which was slightly above the 33·5 new pence for comparable women in all manufacturing industries.

Our conclusions on low pay in the pottery industry are closely in line with those reached by the National Board for Prices and Incomes in their 1970 Report (para. 31):

> It appears then from our survey that there are comparatively few instances of low pay and that these are often associated with short hours. The lowest paid men are not evenly spread through all sectors, but are to be found mostly in the smaller establishments, in earthenware manufacture where the number of small factories is highest, and in certain of the less skilled jobs. For women low pay is again most likely to be found in small establishments, and among makers of domestic ware, but on the whole the lowest paid women are more widely distributed between the various sectors of the industry than their male counterparts. High earnings opportunities appear to exist in certain areas for both men and women. Hours in the industry are not, judged by national averages, generally long for full-time male workers, and most full-time female workers work 40 hours or a little under per week.

JOB SATISFACTION AND WORKING CONDITIONS

During 1970 a sample of women in four firms in the earthenware sector of the pottery industry were interviewed at work. The four factories were fairly representative of the larger earthenware factories in Stoke-on-Trent. In each factory women were selected for interview at random; in all 361 women were interviewed. The aim was to discover why women went out to work and whether or not they were satisfied with their work.† With regard to why women work in the pottery industry, the results of the survey clearly showed that most of the women either had to work to support their families or themselves,

*'New Earnings Survey, 1970, Part 1', *Employment and Productivity Gazette*, (November 1970).

†Christine Lilleker and R. L. Smyth, *Job Satisfaction and Women in the Pottery Industry*, Reports on the British Pottery Industry No. 5, Department of Economics, University of Keele (1971).

or to supplement the family income to an extent that would cause serious inconvenience if they ceased to work. The survey failed to provide support for the commonly held view that women work for pin money or a little extra to spend mainly on luxuries. The women ranked their needs at work on the questionnaire as follows:

1. 'Working with friendly people'   41%
2. 'Adequate wages'   28%
3. 'Having a job which is personally satisfying'   19%
4. 'Security of employment'   11%
5. 'Having a job with a high status'   1%

It is interesting that a social need was ranked higher than wages, which is an economic one. An explanation of this is, as we have seen, that women's wages are adequate in the pottery industry and, in consequence, they were taken for granted. The stress on 'working with friendly people' is an important aspect of working in the pottery industry in Stoke-on-Trent. Very rarely are the relations between the pottery workers and between pottery workers and managers other than friendly. To some extent the friendly relations within the factories reflects the basic friendliness of the citizens of Stoke-on-Trent.

The women at work were asked the following questions: 'Taking everything into account, the pay, the people you work with, the supervision and the work you do, how do you feel about your present job?' The responses to the six possible answers allowed were as follows:

'I love it'   3·6%
'I like it very much'   32·1%
'On the whole I like it'   55·7%
'I'm not very keen on it'   8·0%
'I dislike it a great deal'   0·3%
'I hate it'   0·3%

Thus of the vast majority of the women, 91% stated that they liked their job, whereas only 9% stated that they did not. The same question had been put to women working in electronics factories by a research team at the University of Bradford and their results were significantly different, 79% of their sample were generally satisfied with their job, whereas 21% were not.* In comparison with most other studies that

*A. B. Hill, R. Wilde and C. C. Ridgway, *Women at Work an Investigation into Motivation, Job Satisfaction and Labour Turnover among Workers in the Electronics Industry*, University of Bradford Management Centre (February 1969). Also A. B. Hill, 'Job Satisfaction and Labour Turnover Among Women Workers', *Journal of Management Studies*, Vol. 7, No. 1 (1970).

have been made on job satisfaction, the women in the pottery industry in Stoke-on-Trent appear to be more satisfied with their work than is usual.*

The explanation did not lie in the unduly high proportion of older women in the labour force of the pottery industry. (The older one is and the longer one has held a job the more likely one is to be satisfied with one's job.) It would seem that the social cohesion and relative isolation of Stoke-on-Trent, combined with the peculiar technology of the industry which permits factories to be relatively free of noise and permits workers to set their own pace and talk whilst working, played their part in promoting the high degree of job satisfaction revealed by the survey. The research into job satisfaction was restricted to women; our remarks about the attitudes of men in the pottery industry to their work are based merely on conversations in pubs, at football matches and at W.E.A. and trade union meetings. The men tend to be more critical of their working conditions and managers. They feel that much more needs to be done to provide adequate opportunities for men to change their occupations within the factories. They feel strongly that workers are not encouraged sufficiently to use their initiative or to seek promotion. A number of instances have been pointed out to us of skilled men who keep to their original occupation after it has been, in large measure, de-skilled. Frequently they are given special rates which recognise their special circumstances. Many of the men in this category would welcome an opportunity to learn a new skill. For various reasons managers appear to ignore this source of job dissatisfaction. However, the men do not feel alienated. They identify themselves with their firms and, by and large, they are proud to be pottery workers in spite of there being plenty of room, in their opinion, for improvement. It would seem that men in the pottery industry too enjoy a high degree of job satisfaction.

It was noted in Chapter 1 that working conditions in the pottery industry have improved considerably since 1939. The following comments on the working conditions in the industry in 1970 are reproduced from the National Board for Prices and Incomes Report on the Pottery Industry (Appendix B):

*The National Board for Prices and Incomes reported that according to their survey of workers in selected service industries 'the proportions of workers who said they either liked their job very much or quite a lot was 86% of full-time men, 87% of full-time women and 86% of part-time women workers (*General Problems of Low Pay*, Report No. 169, H.M.S.O., Cmnd. 4648, April 1971). In the Report the following note appears: 'In his book *Alienation and Freedom*, R. Blauner has pointed out that when general job satisfaction questions are asked of workers they almost invariably reply that they like their work.'

Technological advances in the industry have been accompanied by striking improvements in physical conditions. Pottery has traditionally been regarded as a hot and dirty industry, with extremes of heat and cold, and overshadowed by the two classic industrial diseases of pneumoconiosis—'potters rot'—and lead poisoning.

The greatest single contribution to cleaner and more comfortable conditions has been the now universal replacement of the old coal fired bottle ovens by modern oil, gas or electricity fired kilns. This development, together with greater attention to cleanliness and the elimination of dust by observance of the 1950 Pottery (Health and Welfare) Regulations, has transformed the physical surroundings of the industry. Problems of excessive heat in certain areas of the factory, particularly in the summer months, still exist, but managements are generally aware of this problem and do what they can to deal with it.

Accident rates in the industry are rather lower than the average for manufacturing industry as a whole reflecting a relatively low degree of mechanisation, although as a concomitant of this handling injuries—strains and sprains caused by lifting—are more common.

As far as industrial disease is concerned, lead poisoning has been virtually eliminated by changes in the composition of glazes and colours and no case has been reported since 1952. Pneumoconiosis, caused by the dust-laden atmosphere of pre-war factories and by the use of powdered silica for bedding ware in kilns, has proved a more intransigent problem. A major breakthrough was the replacement, completed throughout the industry by 1947, of silica with alumina for bedding, but the problem remained of dust produced in certain making processes, principally towing and fettling. The British Ceramic Research Association (BCRA) has been engaged for twenty years on the problem of dust elimination and has designed several types of extraction hood for the protection of operatives engaged on dusty processes. The adoption of Terylene protective clothing has also been a major step forward. As a result of these efforts and close observance of factory regulations the problem has largely been mastered and the disease is no longer regarded as a hazard, except in a few isolated areas, by either management or operatives.

In 1971 the Second Report of the Joint Standing Committee for the Pottery Industry reported in *Pattern for Progress*:*

---

*Department of Employment, H.M.S.O., p. 10.

Recent developments in ceramic technology and the structure of the industry have engendered a willingness to relinquish the ties of tradition. This attitude to change, as well as the changes themselves, can now be exploited to secure improvements in safety and health.

Further changes in body composition as already facilitated or necessitated by the availability of raw materials and developments in manufacturing techniques are likely to result from the continuing research into methods of production and materials used. Exploitation of these changes as at present foreshadowed by the introduction of reduced silica hotelware body may well lead to an extension of low silica body recipes so resulting in a greater measure of intrinsic safety. Nevertheless, this is a long term project and there will be a continuing need for production methods to be designed to operate safely.

The control of primary dust sources at mechanised operations has enabled some stages in production to be totally enclosed. The future need will be the integration of these separate units into a comprehensive flow line system based on the principle of closed circuit production. Modern tile making plant and large capacity domestic ware production units are already developing in this direction. Such standards of enclosure which would ensure complete control of both primary and secondary dust sources are unlikely to be generally attainable in the more immediate future, particularly in the small and less mechanised factories.

It is recorded in the Report that the number of assessments for pneumoconiosis pensions attributable to the pottery industry in Great Britain fell from 128 in 1950 to 50 in 1960 and 28 in 1969.

LABOUR TURNOVER AND SCHOOL LEAVERS

'The need for a more positive approach to questions of personnel management is, . . . emphasised by the industry's problems in respect of high labour turnover, difficulties in recruitment and absenteeism, which all have adverse effects on costs and productivity.'*

Problems of labour turnover and absenteeism are shared by the pottery industry in common with all other industries. On turnover, the National Board for Prices and Incomes' report revealed a turnover

*National Board for Prices and Incomes, Report No. 149, H.M.S.O., London, para. 75 (July 1970).

rate for males in the year ending January 1970 of 43·2% and for females over the same period a turnover rate of 52·8%. Turnover was defined as the number of adult workers who left their employment between January 1969 and January 1970, expressed as a percentage of the average numbers employed at those dates. The figures are too high and it is understandable that pottery managers are concerned to bring greater stability to their labour forces. Turnover is not spread evenly over all occupations and it varies considerably according to the age and sex of the worker. In spite of the high incidence of turnover it is possible to point to important elements of stability in the relations between employers and workers: 'Over 55% of the men in the pottery industry had 5 or more years' service with their firm . . . nearly a quarter of the men in the industry were reported to have served at least 15 years with their firm . . . the overall proportion of women in the pottery industry with 5 years' service or more was about 46%, some 10 percentage points below the figure for men . . . about 1 in 6 of the women had served for at least 15 years (compared to 1 in 4 of the men).'* It is usual for women with young children to leave the labour force temporarily, and this helps to explain why the length of service of women is below that for men; also why women's turnover rates tend to be higher. Particularly high turnover rates occur among boys and girls who, among other things, 'shop around' a lot and find it difficult to accept the discipline of any factory. Also, labourers and relatively unskilled workers have much higher turnover rates than have skilled workers. As workers get older their desire and ability to change their employer diminishes and turnover rates are much lower for older workers than for younger ones.†

Turnover rates are sensitive to labour scarcity: as unemployment increases so labour turnover tends to fall. From 1945 until 1970 labour was particularly scarce in the Stoke-on-Trent area, the growth of new industries and services providing ample scope for changing jobs. A lot of the job changing took place within the pottery industry as firms succeeded in attracting specialised labour away from neighbouring firms. As a manager in the pottery industry expressed it: 'If it comes through the gates and is still breathing—sign it on'. We cannot say at this stage if mobility occurs because workers are seeking higher earnings or that they simply feel unsettled and desire a change. Clearly

*National Board for Prices and Incomes, Report No. 149, H.M.S.O., London, Appendix A (July 1970).

†Data relating to these relationships in four pottery firms are presented and discussed in *The Worker and the Pottery Industry* by D. L. Gregory and R. L. Smyth, Department of Economics, University of Keele (1971).

with vacancies in every factory there is a temptation to move around, particularly when many persons are already doing it. The National Board for Prices and Incomes in their Report noted that: 'Smaller firms individually were more likely to have a low rate of turnover, but over the whole industry the position was much the same for all sizes of establishment.' This suggests that as the industry has become more concentrated into a smaller number of large firms, and large firms tend to operate larger factories, labour turnover has increased in consequence. We have described how Wedgwood Ltd. has developed its personnel department in recent years and similar developments have occurred in Doulton, Allied English and Royal Worcester. The large groups all recognise that their labour forces deserve very special attention and that care must be taken to provide training, rational wage structures, proper procedures for promotion, etc. by properly ·qualified and experienced staff. As the larger groups and firms obtain economies in marketing, finance and the utilisation of plant and equipment they also face extra costs of sustaining the morale of their labour forces.

Table 10.3, which is based on average turnover figures for the pottery industry and all manufacturing industry for the period 1949–1968, shows that women in the pottery industry have lower turnover rates than women in all other industries in total. In contrast, the turnover rates for men are similar in the pottery industry and in all other industries. (The turnover rates quoted in the text are annual rates whereas the rates in *Table 10.3* are monthly rates and this helps to explain the differences in magnitude between the percentages.) Our

Table 10.3   LABOUR TURNOVER IN THE POTTERY INDUSTRY AND ALL MANUFACTURING INDUSTRIES (1949–1968)

	Engagements: range per 100 employed	Separations: range per 100 employed
Pottery Industry		
Females	2·7—4·8	2·6—4·3
Males	1·9—3·5	1·5—3·0
Total (Males and Females)	2·5—4·3	2·1—3·7
All Manufacturing Industry		
Females	3·0—5·4	3·4—4·9
Males	1·5—3·3	1·7—3·1
Total (Males and Females)	2·0—4·0	2·2—3·7

Source: *Employment and Productivity Gazette* (October figures for each year averaged over the period 1949–1968).

job satisfaction survey findings suggest that some of the labour turnover in the pottery industry occurs in spite of a high degree of job satisfaction and not because there is a high incidence of job dissatisfaction. This suggests that there is a management weakness in the industry as a whole, as the quotation at the beginning of this section states, on the personnel side; for too long pottery firms have regarded labour merely as a service to be bought with money. Fortunately pottery managers have recognised the need in recent years to take positive action on the personnel side. The activities of the Training Board has encouraged these developments.

School leavers constitute a vitally important source of labour to the pottery industry as they do to any industry. Many pottery firms have reported difficulties in recruiting school leavers and in retaining those that had been recruited. Some impression of the magnitude of the problem can be gained from considering the decline in the number of school leavers entering the industry over a decade. In 1958, 21·9% of the boys and 39·8% of the girl school leavers in Stoke-on-Trent who began employment, entered the pottery industry. By 1968 only 11·8% of the boys and 14·2% of the girls entered the industry. Over the same time-period, employment in the industry in Great Britain fell from 68·25 thousand to 61·6 thousand. This contraction would reduce the recruitment of school leavers, but not to the extent that actually occurred.

In part, the competitive position of the pottery industry in the labour market for school leavers in Stoke-on-Trent has been weakened by the development of new industries and services in the area. A study of the employment destination of school leavers in the period 1958–1968 reveals that young persons gravitate towards areas of expanding activity; the growing engineering sector in the area increased its intake of boy school leavers from 17·9% to 35·3% and the employment of girl school leavers in office jobs increased from 16·8% to 40·3%.

Much of the industry's recruitment problem is related to its 'image' in the eyes of the school leavers, teachers and parents. The 'image', or the popular conception of the industry, has to be regarded in some respects as a 'legacy' from the past and in others as self-inflicted. Attitudes appear to be conditioned mainly by stories of what conditions used to be like in the industry. Whilst local prejudices of this sort are perhaps understandable in a close knit community with long standing pottery tradition, the failure to consider the very real improvements which have been made in removing many of the 'hot' and the 'dusty' processes from the modern pottery factory necessarily contributes to the derogatory attitudes which prevail amongst the local community. The industry itself must bear some of the responsibility for failing to better its image. Only a few of the larger Groups attempt to sell themselves to school leavers.

In July 1969 a survey of secondary school leavers was undertaken in Stoke-on-Trent* to determine the nature of the pottery industry's image and compare it with the children's attitude towards seven other local industries. The general results of the survey confirmed that the industry does have a very bad image with school leavers and has a low preference rating as a future employer compared with some of the other industries cited in the survey. To test the children's attitude to local industry a questionnaire was used which contained a simplified semantic technique, whereby the children marked appropriately a three-point scale between seven uniform sets of bi-polar adjectives which described certain aspects of local industries. *Table 10.4* shows the results for the pottery industry.

Table 10.4 SCHOOL LEAVERS ATTITUDE TOWARDS THE POTTERY INDUSTRY

Attitude		Neutral or no opinion (%)		Attitude	No information given (%)
Lively	19·7	18·1	57·6	Dull	4·5
No prospects	46·8	25·0	22·9	Prospects	5·3
Modern	29·0	22·7	41·9	Ancient	6·3
Uninteresting	47·2	15·7	32·7	Interesting	4·4
Soft	25·2	30·3	38·2	Hard	6·3
Bad pay	26·4	29·2	40·1	Good pay	4·2
Good	21·0	31·3	42·1	Bad	5·6

The data shows that the children in the sample regarded the pottery industry as 'dull', 'uninteresting', 'lacking in prospects' and 'ancient' rather than 'modern'. It is interesting to note that the majority felt the industry gave 'good pay' rather than 'bad pay', despite the local belief that the industry is generally underpaid. In fact, the earnings of young workers in the pottery industry do not appear to differ significantly from their counterparts in other manufacturing industries (*Table 10.5*).

In recent years the contracting supply of school leavers to the industry has not had severe effects on the labour force. This has been so, principally for two reasons. Firstly, the overall decline in the industry's demand for labour in the last 10 years; secondly, the fact that many

*The results of this survey are analysed in greater detail in *The Worker and the Pottery Industry* by D. L. Gregory and R. L. Smyth, Department of Economics, University of Keele (1971).

Table 10.5 EARNINGS IN POTTERY AND OTHER MANUFACTURING INDUSTRIES

Industry	Average hours worked		Average hourly earnings (pence)		Average weekly earnings (s. d.)	
	Youths & Boys (under 21 years)	Girls (under 18 years)	Youths & Boys (under 21 years)	Girls (under 18 years)	Youths & Boys (under 21 years)	Girls (under 18 years)
Pottery	41·2	38·7	68·5	48·4	235 1	156 0
All manufacturing industries	41·2	38·5	68·6	50·3	235 6	161 6

Source: *Employment and Productivity Gazette* (February 1970).

smaller pottery firms have recruited older and married women by offering them special accommodating hours of work. It is probable that the industry's labour force will, with further concentration of productive capacity, continue to contract in the near future, and that this will offset the decline in the recruitment of school leavers. However, there are strong indications that the industry will exhibit an increasing 'replacement demand' for labour in the future. The Prices and Incomes Board noted that 'the average age of operatives in many establishments is high. Often the bulk of the labour force may be made up of long service pottery workers in their 40s and 50s . . .' (para. 19). It would appear that many firms with 'ageing' work forces will be faced with replacement problems. This need could be partially satisfied by greater recruitment of school leavers. The fact remains that for the industry to be successful in this objective a greater collective effort will have to be made to convince the school leaver that the industry does offer 'prospects' and is *not* 'uninteresting' and 'dull'.

INDUSTRIAL RELATIONS

The pottery industry has long had an excellent record of industrial relations. There has been an almost complete absence of strikes and few disputes that have lasted for more than a day or so. The industry possesses a distinctive institution of an annual settling time which originated in the 1890s. Perhaps the absence of strikes and the annual settlement are connected. We will return to this possibility after describing how the two sides of the industry are organised and looking at methods of wage determination and plant as well as industry rate fixing procedures.

THE BRITISH CERAMIC MANUFACTURERS' FEDERATION (BCMF)

The British Ceramic Manufacturers' Federation has been described in Chapter 3. Its Wages Advisory Committee, which contains representatives from all sectors of the industry, conducts the annual negotiations on the National Joint Council (NJC) with the Ceramic and Allied Trades Union, the only union represented on the NJC. The NJC was reconstructed in its present form in 1945.

THE CERAMIC AND ALLIED TRADES UNION (CATU)

About the same time as the BCMF was founded, the many potters' unions amalgamated and formed the National Society of Pottery Workers; in 1970 the name was changed to the Ceramic and Allied

Table 10.6 MEMBERSHIP OF THE CERAMIC AND ALLIED TRADES UNION 1962–1971

Year	Males	Increase over previous year (%)	Females	Increases (%)	Total*	Increase over previous year (%)
1962	9 960	—	12 269	—	22 229	—
1963	10 428	4·7	12 295	0·2	22 723	2·2
1964	11 417	9·4	12 932	5·2	24 349	7·1
1965	11 804	3·4	13 944	7·8	25 748	5·7
1966	12 232	3·6	15 366	10·2	27 598	7·2
1967	12 694	3·8	16 233	5·6	28 927	4·8
1968	12 952	2·0	16 625	2·4	29 742	2·8
1969	13 436	3·7	16 882	1·5	30 361	2·1
1970	15 056	12·0	18 160	7·6	33 227	9·4
1971	16 921	12·4	18 894	4·0	35 815	7·8

Source: CATU.
*Total includes juveniles 1968 and after.

Trades Union. Most pottery workers who are union members belong to CATU. The Transport and General Workers Union are responsible for a few transport workers employed by pottery firms and maintenance workers are members of electrical and engineering unions. Unions other than the CATU tend to keep their rates in line with what is agreed each year by the National Joint Council. In spite of a fall in the labour force of the pottery industry the CATU has increased its membership; this is shown in *Table 10.6*. *Figure 10.2* shows the internal organisation of the Union.

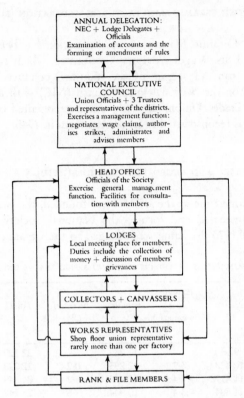

Figure 10.2   Lines of communication within the structure of the Ceramic and Allied Trades Union

THE NATIONAL JOINT COUNCIL AND THE ANNUAL SETTLEMENT*

The National Joint Council for the Pottery Industry is composed of 12 representatives of the BCMF and 12 of the CATU; the chairmanship alternates between the two sides. The aim is to concentrate negotiations at national level and provide 'a coherent, minimum rates structure with uniform overtime, holiday and related provisions'. The occupational groups and basic rates and conditions are set out in the Pottery Industry Wages Structure Agreement . . . A particular feature of the NJC agreement is that industry-wide, as well as departmental (i.e. sector) and individual (i.e. plant level) alterations to

*The quotations in this section are taken from the National Board for Prices and Incomes Report (No. 149) on the Pottery Industry.

established rates, can be made only on a fixed annual basis unless there is agreement between the management and the operatives concerned to do otherwise . . . any proposals to change existing terms, whether by the employers or the trade union, must be exchanged between the parties in the form of written notices within six weeks prior to 25 March of any year. . . . The annual round of industry-wide negotiations begins in the autumn with consultations undertaken by both sides with their respective memberships in order to gather views and formulate proposals. The resulting notices are by tradition formally exchanged at 5 p.m. on the due date. The trade union's claim is then costed by certain member companies of the BCMF in order to determine the average cost of granting the wage and other improvements. Negotiations continue until a settlement is reached, which is then submitted for simultaneous approval by both sides' memberships and is normally effective from 25 March. During this period, individual notices are also negotiated at plant level between management and the operatives concerned. The National Board for Prices and Incomes noted the movement of basic time rates paid by the great majority of firms were usually in line with the increases negotiated in the NJC. However it was noted that the NJC does not negotiate piece rates: 'These are negotiated with local managements for each new job. As a result many firms pay above the NJC minima in order to attract labour.' The Board's earnings survey revealed that 'two out of every three time-workers employed by the NJC firms were paid above the NJC rates'. The Board's case studies indicated that 'the NJC wages structure does not, in practice, directly determine the rates for more than a minority of employees, and suggests that the extent of wage determination at plant level even for time-workers is high, with many employers paying above the NJC rates in order to recruit or retain particular categories of manpower. Although the NJC agreement lays down a general standard for the determination of piece-rates, these also, along with payment-by-results bonuses, are essentially determined at local level by individual or group bargaining'. The Board also reported that 'nearly 54% of the total adult labour force (just under 46% of men and 61% of women) were paid by a system related in some way to performance'.

The following passage is reproduced from the *Royal Commission on Trade Unions and Employers Associations 1965–1968* (The Donovan Report)* from which it would appear that industrial relations in the pottery industry are in line with the Commission's conclusions:

*H.M.S.O., Cmnd. 3623 (June 1968).

Britain has two systems of industrial relations. One is the formal system embodied in the official institutions. The other is the informal system created by the actual behaviour of trade unions and employers' associations, of managers, shop stewards and workers. . . . The keystone of the formal system is the industry-wide collective agreement, in which are supposed to be settled pay, hours of work and other conditions of employment appropriate to regulation by agreement. . . . The informal system is often at odds with the formal system. Actual earnings have moved apart from the rates laid down in industry-wide agreements; the three major elements in the 'gap' are piecework or incentive earnings, company or factory additions to basic rates, and overtime earnings. . . . The bargaining which takes place within factories is largely outside the control of employers' associations and trade unions. It usually takes place piece-meal and results in competitive sectional wage adjustments and chaotic pay structures.

Until such time as a comprehensive job evaluation exercise has been applied to the pottery industry's wage structure it will remain if not 'chaotic' at least riddled with anomalies. In the pottery industry the shop stewards or works representatives are less powerful than they are in engineering, mainly because union full-time officials are readily available to help in any shop-floor conflict that arises. The shop-floor bargaining on piece-rates and bonuses to be added to time-rates arise because of the great variations between factories in terms of products and processes in the pottery industry. They are not a symptom of militancy on the shop floor. We may conclude that the annual settlement cannot by itself explain the absence of strikes; the explanation is to be found in the highly localised nature of the industry and the patience and good sense of union members and officials and some employers, combined with the great advantage of an absence of demarcation disputes through there being only one union involved. The absence of strikes in the past means that there is an absence of 'strike culture' among pottery workers in the present.

WAGE DETERMINATION

Of course, wages are not determined by collective bargaining: they are determined by supply and demand forces, or more specifically, by the need to pay sufficient to attract workers away from other industries. Unless adequate wages were paid labour could not be

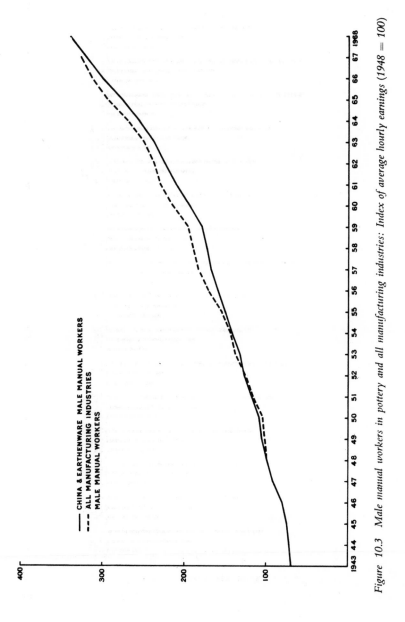

CHINA & EARTHENWARE MALE MANUAL WORKERS
ALL MANUFACTURING INDUSTRIES
MALE MANUAL WORKERS

Figure 10.3 Male manual workers in pottery and all manufacturing industries: Index of average hourly earnings (1948 = 100)

226

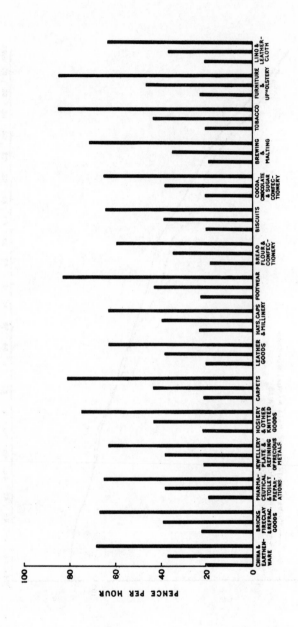

Figure 10.4 Average hourly earnings of full-time female manual workers in selected manufacturing industries in 1948, 1959 and 1968

recruited in required amounts, and the pottery industry is not one in which labour is likely to be paid more than is strictly necessary to get the work done. *Figure 10.3* shows how the average hourly earnings of male manual workers in pottery has risen closely in line with average hourly earnings in all manufacturing industries since 1948. Women pottery workers' hourly earnings behaved similarly over the period. The figure displays the result of informal as well as the formal bargaining. To be more precise, the supply and demand forces (productivity governing the demand side) normally determine the limits within which wages may lie and the sociological force (collective bargaining) determine the actual wages within the range permitted by the economic forces. *Figure 10.4* shows how the earnings of women in the pottery industry have kept in line with the earnings of women in comparable industries.

PRODUCTIVITY

Productivity is not something that can be seen and experienced like output or exports; it is a relation between output and a selected input, the input most usually selected being labour. Although it is a somewhat elusive concept it lies at the heart of the economic performance of the pottery industry. What productivity is and how it changes depends on every aspect of the industry considered in this book and others too. It depends on the number and size of firms: we believe it has improved in the pottery industry as a consequence of take-overs and amalgamations. It depends on exports: sales in the United States and Canada permit long runs of some patterns to be manufactured and as cost per piece falls so productivity increases. It depends on manufacturing standards and on design; also on the value of sales, which in turn depend on merchandising and pricing—elements which are not always explicitly considered when productivity is discussed. Productivity depends on the quality of the work force and the management, and it can even increase as a result of strong trade union pressure: a sudden sharp increase in money wage-rates may stimulate more efficient operations. Conversely, the introduction of a better machine by management will usually lead to a demand for more wages by labour on the grounds that the productivity of labour has increased. In sum we can say that productivity is all-embracing and therefore to argue or believe that too much fuss is made about it is to commit a gross error of judgment.

Figure 10.5 presents a crude measure of productivity movements in

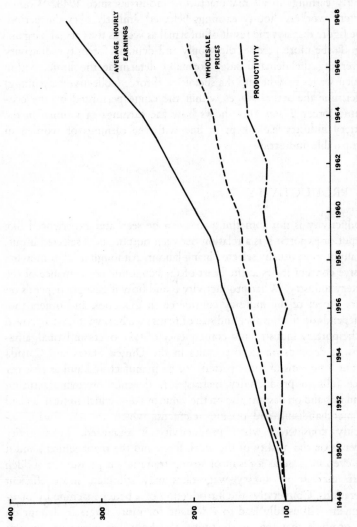

Figure 10.5 Productivity, prices and average hourly earnings in the pottery industry, 1948–1968

the pottery industry as a whole since 1948. It was calculated by dividing an index of employment into an index of physical output for each year. *Figure 10.5* indicates that over the 20 year period productivity increased by 47%. It shows that, over the same period, average hourly earnings for manual workers in the industry increased faster than did physical productivity. For this to happen prices too must have increased, and the increase in wholesale prices for the period is also displayed in the figure. (The price index refers to domestic pottery only so it is not a precise reflection of all pottery prices.)

The *Census of Production* enables productivity measurements to be made for particular sectors of the industry. We require a measure of net output for more precise productivity measurements. For this we use the value added by the industry—the fund out of which wage and salaries are paid and profits taken—which excludes costs of fuels, raw materials and components to which value is added by production. Unfortunately, at the time of writing a census of production for Great Britain has not been published since 1963, so our detailed productivity measurements are restricted to 1963 and the previous Census years, 1948 and 1958.

Productivity is something that is best looked at over time; it is difficult to gain an insight into productivity measured at only one point in time. In addition it is desirable that productivity comparisons be made with comparable industries. *Figure 10.6* presents a measure of labour productivity in the domestic sector of the pottery industry for 1948, 1958 and 1963 and comparisons are made with 15 other consumer goods industries. *Figure 10.7* presents similar data for electrical porcelain, tiles and sanitary ware and compares them with 10 industries which are in some ways similar. The measure used throughout is net output per pound sterling of wages and salaries. Wages and salaries provide a comprehensive measure of labour input and take into account differences in skills and intensity and duration of work as well as the age and sex of workers (to divide simply by number of workers in making comparisons can be misleading). The measure may be thought of as being composed of two components: (1) net output $\div$ number of hours worked, and (2) number of hours worked $\div$ wages and salaries. Of these, (1) will increase when production increases are achieved by means of improvements in technology, improved working methods, marketing, etc.; (2) however, may decrease because the amount of labour that may be purchased per pound is falling. The overall ratio is the outcome of the two conflicting forces. *Figure 10.6*, which relates to the domestic sector of the pottery industry and comparable consumer goods industries, shows that labour productivity in pottery

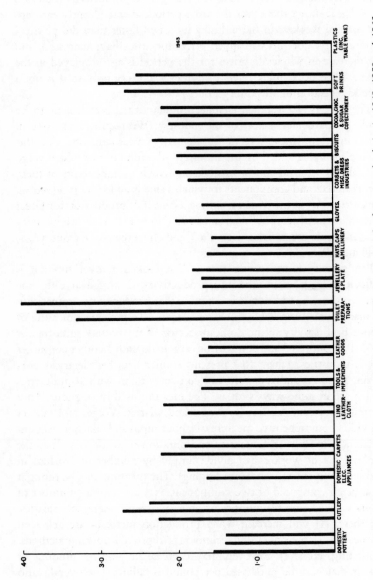

Figure 10.6 Productivity in consumer goods industries: Net output per £ of wages and salaries in 1948, 1958 and 1963

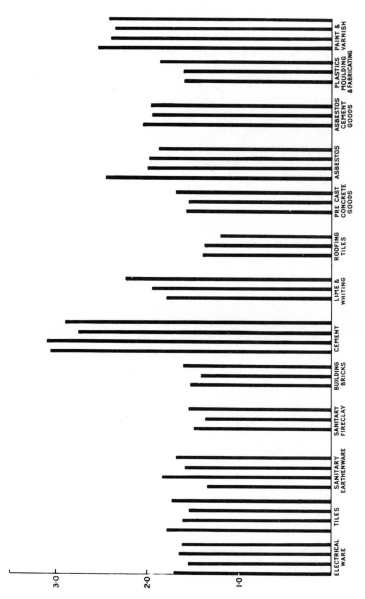

Figure 10.7 Productivity in the building and construction industries: Net output per £ of wages and salaries in 1948, 1954, 1958 and 1963

is particularly low. This could be so because the domestic sector of the pottery industry is less capital intensive than the other industries (a high degree of capital intensity probably explains why labour productivity is so much higher in toilet preparations and soft drinks). As the drive towards mechanisation in the pottery industry increases then, presumably, labour productivity will increase; a low degree of capital intensity can be appropriate if the labour is utilised to produce a wide range of fashion goods for a fickle public. So long as the products may be sold at a profit then a labour intensive industry makes good sense. However, if wages and salaries are increasing and demand is sensitive to price changes then a lack of adquate mechanisation could be inappropriate. *Figure 10.6* also shows that over the period the amount of net output per pound of wages and salaries in the domestic sector of the pottery industry fell slightly; this may be interpreted as wage and salary payments increasing faster than productivity: in other words, a profits squeeze. Given that many firms went out of business or were taken over by other firms in the early 1960s, this would appear to be a plausible explanation of the movements shown in the chart.

Comparison of *Figures 10.6* and *10.7* shows that labour productivity was higher in tiles, sanitary ware and electrical porcelain than in domestic ware; this was to be expected, as domestic ware in 1963 was the least mechanised sector of the industry. It may be seen that, unlike domestic ware, there were rises in the ratio of net output to wages and salaries in most sectors of the industry between 1958 and 1963. It can be deduced from the figures that the domestic sector of the industry experienced the most difficulty in increasing productivity over the period studied. It is reasonable to suppose that the situation has not changed over the period 1963–1971, except in one important respect—the domestic sector has increased productivity to keep in line with increases in wages and salary payments.

Part V

CONCLUSIONS AND FUTURE PROSPECTS

11

CONCLUSIONS AND FUTURE PROSPECTS

There have been rapid developments in the British pottery industry since we commenced writing this book in 1970. The industry has had to endure a long period of stagnation after having enjoyed a prolonged period of excess demand. In the domestic sector of the industry there was an amalgamation between two really large Groups, Doulton and Allied English, and there were strong indications that a new grouping of large earthenware firms would be formed. In tiles, Pilkington became part of the Tilling Group and in sanitary ware Twyfords was acquired by Reed International. Mechanisation and semi-automation finally secured a firm grip on all the sectors of the industry.

In 1972, an engineer in Enoch Wedgwood Ltd. explained that if only five years previously someone had claimed that in 1972 he would be operating a self-loading kiln in Tunstall he would not have believed them. In 1972 his self-loading kiln was operating to his satisfaction. A production manager in the same firm confessed that his battery of making machines now produces better cups, plates and bowls than he produced by hand in 1962. When the Semart Importing Company, an American corporation, acquired Enoch Wedgwood (Tunstall) Ltd. (and Crown Staffordshire China Co. Ltd.) in 1964, their first action was to instal new cloak rooms and restaurants; after that plans were drawn-up for new machines and buildings. Although the work in the factories is now less-skilled and more monotonous, the workers are probably more satisfied with their wages and their working conditions than they were before their work rooms were reconstructed. The experience of Enoch Wedgwood has been repeated in many pottery firms over the past few years.

EMPLOYMENT

The labour force in the pottery industry fell by 28 000 persons between 1951 and 1971. It is likely to fall further in spite of expected modest increases in the volume of output. What happened in the tiles sector of the industry in the 1960s is likely to happen in the domestic sector in the 1970s. A transformation of tile production was caused by mechanisation, and the demand for labour fell substantially as output increased considerably. The regrouping of the domestic sector of the industry into large units and the entry of non-pottery firms ensures that the finance for further mechanisation is available. Data presented in Chapter 6 (on costs) indicated that the pottery industry is relatively more labour intensive than other manufacturing industries. This means that in a period of sharply-rising wages the profits of pottery firms would be particularly vulnerable. They will, in the interests of the shareholders, aim to become more capital intensive, and the demand for labour will fall. Recent developments in technology provide manufacturers with the new machines, kilns and equipment which will render production less labour intensive. The period of 'wait and see' is over with regard to mechanisation; the new machines and processes have been tested and found to be satisfactory. Mechanisation in the pottery industry will increase substantially in the years ahead. Inevitably employment will fall.

A decline in employment in the pottery industry will have serious consequences for North Staffordshire. It is an area which, in the post-war years and up until 1970, enjoyed full employment. However, it was an area from which labour could move easily to find jobs elsewhere. It is highly desirable that new industry be introduced into North Staffordshire to absorb the labour that will be released by the pottery industry.

It was noted in Chapter 10 that the industry in the 1960s failed to recruit an adequate number of juveniles. Perhaps the fall in total employment prevents this weakness from causing harm to the industry. Unfortunately, an industry with a labour force with a higher-than-average proportion of elderly workers can be adversely affected by a lack of adequate recruitment in spite of a fall in total employment.

Many of the highly-skilled workers in the industry are elderly and their skills will still be required when more machines have been introduced. The industry will need to attract more school leavers in the years ahead to train for the skilled jobs which will remain in an industry which will be more capital intensive than at present.

By and large the new machines will not replace highly skilled workers. It is likely that they will enable skilled workers to use their special skills in a more sustained fashion than in the past. As layout has improved in pottery factories, the carrying and fetching and placing of ware has been more and more performed by fork-lift trucks and by conveyors rather than by skilled workers and labourers. It is likely that a number of tasks will be performed more economically by hand than by automated machines in the future. Thus only some ranges of pottery will be produced by plants which are fully automated and the most likely developments will be semi-automated plants. The market is such that frequent changes in shapes and decorations are required, and this tends to render fully-automated plants uneconomic. For many production tasks in the pottery industry it will not make sense to attempt to substitute highly complicated machines, which are expensive, for 'highly complicated' workers who, in comparison, are relatively cheap. However, these aspects of mechanisation and automation will not prevent the total labour force from falling substantially.

SUBSTITUTE MATERIALS

How do we know that a new and reasonably cheap new material will not replace clay in earthenware and bone china manufacture before 1980? Why is it most unlikely that the pottery industry will become a sub-division of the plastics and glass industries?

First of all, pottery raw materials are abundant and cheap. The vast quantities of clays, flints and sands that exist render it highly unlikely that the natural materials will give way to synthetics on a substantial scale. The maintenance of adequate supplies of bone has worried bone china manufacturers in the past 10 years. However, it is reasonable to assume that supply will react favourably over the long run to price increases. Secondly, the many improvements which have been made since 1948 in making and firing should enable manufacturers to sustain output by keeping costs and prices in line with those of toughened glass and plastics. The linen and cotton industries are examples of economic activities based on natural raw materials which have, in large measure, been replaced by synthetics, rayon and terylene. It does not follow that pottery too must decline in a similar manner; glass and plastics appear to be less satisfactory substitutes for pottery than are synthetic fibres for cotton and linen. Thirdly, British manufacturers of bone china are well-established on world markets and foreign countries could not readily make it for themselves. Bone china, both from technical and aesthetic points of view, is an out-standingly good material. There is a big potential for increasing the output of bone china, and only partially at the expense of earthenware. British manufacturers, because of bone china, have an advantage over rival firms in Japan and Germany, and by exploiting it they should succeed in holding their own on the world market. One cannot, however, be too optimistic about British exports in the long run. Red China may fairly quickly develop into a formidable competitor and, as was only to be expected, Japan has commenced to manu-facture bone china.

One important factor which may reduce the size of pottery exports from Britain is the availability on world markets of materials process-ing and making and firing equipment, which renders it easier than it has been for foreign countries to establish their own pottery industries.

DEMAND STABILITY

We have given reasons why the pottery industry is likely to hold its shares of the domestic ware, tiles, sanitary ware and electrical ware markets in the years ahead. It has been noted that both economically and technically ceramics is a tough competitor from the point of view of manufacturers of substitute materials. However, there is a possibility that the total market will contract because of a change in consumer tastes. Perhaps sales will fall because British pottery manu-facturers merely succeed in holding on to a constant share of a shrinking market.

The demand for industrial ceramics is derived from activity in building and in electricity generation and distribution. Building activity will expand considerably in the future, which means that producers of tiles and sanitary ware can look forward to expanding markets. Electrical ware will obtain a substantial boost once it is decided in Britain to erect another super-grid. This, however, may not happen this century. Demand will increase once prosperity returns to the heavy electrical engineering industry. In 1972 the industrial ceramics sector of the industry, excluding electrical ware, tiles and sanitary ware, remains small and this may be the beginning of something big.

What about the biggest sector of the industry? Will the demand for tea, coffee and dinner sets and ornamental ware be sustained? At present, tableware producers are highly dependent on the tradition of gift-giving to the bride and groom by relatives and friends. Only a few young married couples purchase their own dinner, coffee or tea sets. There are indications that this important source of demand for domestic ware will decline in the future. Weddings are either being dispensed with altogether or are becoming less formal. The consequence is that the demand for wedding presents is likely to decline. In the future fewer families are likely to possess unused china. Although the wedding present demand is likely to decline, however, it is reasonable to suppose that there will be compensations. Wedding presents will continue to be given and received and they will take a different form. There will be a tendency to substitute coffee mugs and soup bowls for tea sets. Also the families who do not receive wedding presents are likely to purchase tableware on their own account.* Although afternoon tea parties are not what they used to be and the pattern of demand for tableware is changing, the end result need not be a fall in the demand for domestic pottery. On the credit side, income and leisure are increasing and these two factors are likely to sustain the demand for formal tableware. Also eating-out is on the increase, and this will increase the demand for hotel ware. We noted that this sector of the industry had been strengthened considerably in recent years.

The demand for ornamental ware, the ugly and the artistic, gives every indication of increasing as income per head increases. More income and leisure encourages gift-giving and the pottery industry is properly organised to cater for this growing demand.

Finally, our 1969 Tableware Survey† revealed very wide variations between families at similar levels of income in their ownership and

*In 1971 Taylor & Kent Ltd., fine bone china manufacturers, Stoke-on-Trent, conducted a market survey. The following passage is taken from the conclusion: 'It would appear that the increase in the use of convenience foods and instant drinks has brought with it a corresponding decline in popularity of certain tableware and an increase, in the popularity of multi-purpose vessels, mainly the coffee mug. In households where both the husband and wife are at work all day there is a tendency to cut down on unnecessary labour, thus not using a full dinner set as washing up dishes encroaches on their leisure time.' The survey was presented by A. Kusmirek of the marketing department and Miss J. Pearson his assistant.

†P. W. Gay and R. L. Smyth, *The Household Demand for Tableware*, Department of Economics, University of Keele (1970).

purchasing of pottery. Many families purchase very little pottery indeed, year after year. The substantial gaps that exist on the home market, given imaginative and aggressive marketing, could gradually be filled. As more and more families are stimulated to purchase, demand will be raised permanently to a higher level. Evidence has been presented which suggests that expenditure on pottery tends to generate more expenditure, not less. In this respect pottery is like foreign travel or alcohol. The amount spent on pottery and glass per family has been shown to be low and there appears to be plenty of scope for improvement. On balance we have good grounds for believing that the market for domestic pottery will continue to increase in the forseeable future.

The following proposition has been put to us more than once: 'Surely the British pottery industry must collapse soon. Except for fine bone china, it is being squeezed out of foreign markets by a combination of tariffs and Japan. At home plastics and toughened glass appeal to more and more consumers because they truly reflect contemporary taste. Pottery, in contrast, is old-fashioned and steadily it must lose its grip on the mass market.' We have already argued that the latter proposition is fallacious: the remarks apply only to a relatively small sector of the domestic market and they are not relevant to a vast range of industrial ceramics. In practice, modern technology is generating new demands for ceramic materials which can resist newly-developed highly-corrosive liquids and solids, and industrial processes which require moulds and tools which can resist high and sudden changes in temperatures. The first proposition overlooks the strength of British manufacturers on world markets in medium and low priced earthenware and bone china; and this is in addition to the entrenched positions enjoyed by British manufacturers of fine bone china, particularly in the United States and Canada. Even if some sales are lost in North America there are opportunities ahead in Europe. Also, tableware and ornamental ware depends very much for its sales on individuality in design and texture, and foreign customers prefer to purchase traditional British designs and shapes made in Britain and not in Germany or Japan. We can find no support whatsoever for the proposition that the British Pottery Industry is in danger of imminent collapse.

INDUSTRIAL RELATIONS

Relations between labour and capital in the pottery industry were presented in Chapter 10 in terms of co-operation rather than conflict.

The combination of an annual settlement and the concentration of the industry in North Staffordshire have contributed towards good industrial relations. Equally important has been the North Staffordshire characteristic of, by and large, being content with one's lot. Shop stewards, mainly because of the ease of access of managers to full-time union officials, are not nearly so dominant as they are in many larger industries. It is likely that satisfactory industrial relations will persist in the industry. However, this will not just happen: it is something that will have to be worked for. Fortunately the large groups recognise that labour is a key factor in their plans for profits and growth. They have abandoned the traditional 'take-it-or-leave-it' of pottery employers towards workers. Personnel and training officers are now firmly established in the industry. In smaller firms, relations between employers and employees continue to be reasonably satisfactory. The intimate daily contacts appear to compensate for any feelings of dislike and distrust that happen to be present.

OVERSEAS BRANCHES OF BRITISH FIRMS

It has been noted at various points in the text that British pottery firms have established factories abroad. Firms usually appoint agents in the first instance, then if exports grow they may establish their own showrooms and warehouses. The last step, and one only occasionally taken, is making ware abroad. The main reason for manufacturing abroad appears to be the desire not to be shut-out of large markets by tariffs and quotas. In countries in Africa and Asia British firms have been invited to establish factories by governments and by the Ministry of Overseas Development. With some firms, Doulton & Co. Ltd. and H. R. Johnson–Richards Tiles Ltd. are examples, it would appear that there was a strong desire to become international companies as part of their growth strategies. H. R. Johnson–Richard Tiles Ltd. produce identical tiles in various countries and by using standardised packaging they can supply any one market from a number of countries.

Changes in technology and improvements in organisation and marketing by suppliers of raw materials and machines and equipment, by British, French, Italian, Japanese and German firms, are rendering it easier for firms to be established abroad in developing as well as in developed countries. Because of the improvements in technology, which are built-in to the machines and equipment and kilns and buildings readily available on the market, only relatively few skilled workers are required to operate the new factories. It is easier to train and direct machine-minders than potters. Most overseas subsidiaries make sanitary ware, tiles and electrical ware. The new machines and equipment render 'it likely that similar developments will occur in domestic ware and not only in the cheaper ware. In addition, countries abroad will develop their own factories and not rely on attracting subsidiaries of British firms.

These developments, which accelerated considerably in the 1960s, could have serious consequences in the long-run for British exporters. However, there is no need to be unduly pessimistic about this, for as countries become more industrialised the desire for a wider range of industrial and domestic ceramics will develop and, more important, the ability to pay for imports will also develop.

COMPETITION AND MONOPOLY

Writing in 1955, the United States Attorney General's Committee on Anti-Trust Laws considered that three factors were extremely important for 'effective' or 'workable' competition to be present in an industry. They were:

1. The number and size distribution of firms must guarantee that no one firm or group of firms acting in concert could hold for long the power to choose its level of profits by giving less and charging more.
2. Relative freedom of opportunity for entry of new rivals and the cost of entry should not be impracticably high.
3. There should be genuine independence on the part of the business units in an industry.*

The pottery industry in 1972 clearly satisfies the three main criteria for workable competition to exist. It is a small industry with some 100 firms operating in it. The various business units are genuinely independent. Raw materials, machinery and equipment are freely available to anyone with money to spend and, in consequence, entry is not difficult. New entry by other than very small firms rarely occurs; this is because profits in the industry are modest and have to be worked for, not because there are restrictions on entry. Over the years the

*Alex Hunter (Ed.), *Monopoly and Competition: Selected Readings*, Essay 5, Penguin, London (1969).

government has left the industry alone, and for good reason; the firms in it are disciplined by competition, and Government regulations, in addition, are not required to achieve social ends. The Attorney General's Committee noted the possibility that small firms may be tolerated by large firms to provide an illusion that genuine competition based on numbers exists. In the pottery industry, however, small firms compete with a fair degree of success with the larger ones in the domestic ware sector. They are not kept in business to camouflage monopolistic practices.

The tiles sector of the industry is dominated by H. & R. Johnson–Richards Tiles Ltd., although stiff competition is provided by Pilkington Tiles and others. Nevertheless, the prices charged are low and the price-fixing arrangements which exist to encourage standardisation and enable firms to achieve economies of large-scale production have had the blessing of the Restrictive Practices Court. Competition is extremely severe in sanitary ware where mergers have resulted in four large firms accounting for most of the output. Electrical ware production is dominated by one large firm, Allied Insulators Ltd., which is not in a position to raise prices unduly, even if it wanted to, as there is only one main source of demand—the various Electricity Generating Boards. In spite of the greater degrees of concentration in industrial ceramics, the only danger of monopoly lurks in the domestic sector. There are three large groups in the domestic sector: Royal Worcester, Wedgwood and Royal Doulton Tableware, and if further concentration occurred then an unsatisfactory situation could arise. The danger is merely a potential one. Further substantial takeovers by one of the 'Big Three', or a merger between any two of them, could result in retailers having their choice of suppliers unduly restricted. However, it is not really worth worrying about a hypothetical situation as at present the competition on the home market between rival suppliers is sufficiently severe to please either Enoch Powell or Adam Smith.

MERGERS

Our conclusion on mergers is that in the 1960s they were long overdue and that the increase in size of the administrative units which resulted was entirely appropriate to changing business conditions. They resulted in more money being injected into an industry that was operating with inadequate and run-down capital. Mergers permitted management to improve considerably and, most important, they enabled

British firms to compete in terms of equality with German and Japanese firms. In some cases, after merger, individual units operated very much as before. This suggests that some mergers were not really necessary. They seemed to have been triggered-off on the flimsy grounds that 'it pays to be bigger'. We know of only a few mergers in the pottery industry to which these remarks apply and the reason may be merely a matter of timing. Perhaps it is unreasonable to expect new policies to be implemented immediately after mergers occur, at present we may be too near events to attempt judgment.

An interesting aspect of the mergers which have occurred is that pottery firms have, with few exceptions, acquired other pottery units. They have refrained from diversifying into, say, cutlery, hardware or general engineering. The move into electronics by The Royal Worcester Porcelain Company was exceptional. In contrast, Doulton became more specialised when it disposed of its interests in glazed pipes. It was outsiders coming into the industry by means of takeovers which provided an important element of diversification in the industry. The Pearson Group, which controls Royal Doulton Tableware Ltd., is highly diversified, as is Carborundum Ltd. which controls Spode Ltd., and the Semart Importing Co., which controls Enoch Wedgwood (Tunstall) Ltd. and Crown Staffordshire China Co. Ltd.* The access to outside finance and management expertise which resulted when outsiders diversified into the pottery industry must be regarded as a source of strength and a welcome development.

SMALL FIRMS

The pottery industry provides evidence of take-over bids that have been both constructive and beneficial. It also provides excellent examples of small firms which have survived and are flourishing. They provide consumers with an extended range of choice and, frequently, they can indulge in price-cutting which, whilst it annoys larger firms, can be a boon to consumers. They also provide for managers and workers who would be less contented if they worked in larger units. Their presence in the industry prevents larger firms from being unduly complacent. Frequently they provide services and specialised products which cannot readily be provided by large firms.

In Chapter 5, the activities of Elijah Cotton Ltd., T. G. Green Ltd.,

*Acquired by Wedgwood Ltd. in 1973.

Thomas Poole & Gladstone Ltd., Portmerion Potteries Ltd and Arthur Wood & Sons (Longport) Ltd., were described, and George Wade & Son Ltd. was considered in Chapter 9. They are all relatively small firms which do not find that their size is a disadvantage in earning profits. Indeed, they would claim that their larger rivals suffer from inflexibility and unduly high costs of administration. Many other successful small firms operate in the domestic sector of the industry they include: A. G. Richardson & Co. Ltd., an earthenware firm in Stoke-on-Trent which, in 1972, acquired Poultney & Co. Ltd., a firm which also produces earthenware in a newly-built and elegant factory in Cornwall; Biltons (1912) Ltd., a competently managed firm which competes successfully with, among others, the Clough Group and Staffordshire Potteries Ltd; Simpsons (Potters) Ltd., whose ware is familiar to Dukes and Duchesses, was acquired by an American firm in 1971 (the late Managing Director, Mr. T. Simpson, explained to us that only his firm made really good dinner sets). Away from North Staffordshire there are Hornsea Pottery Co. Ltd. and Govancroft Potteries Ltd., Glasgow.

It has been suggested that the small firms will in the not too distant future be either squeezed out of business or taken-over. It is most unlikely that things will work-out this way. Although the industry will be dominated by a small number of large firms, nevertheless there will be ample scope for small firms to remain in business. A few of them will, presumably, become established as large firms. Small firms attract employees and customers who are not attracted towards large groups; also governments now consider that efforts should be made to ensure that small competently managed firms should flourish.

PAST, PRESENT AND FUTURE

It is the contrast between the British pottery industry as it existed before 1939 and as it is at present that is so interesting. Pre-war the industry was peculiar. Like the fishing industry or the docks, it was dominated by tradition, peculiar working practices abounded and it lacked a sound technology based on scientific investigations. There were far too many small firms. Profit rates were miserably low and, like textiles, coal and shipbuilding it was one of Britain's traditional depressed industries. The manufacturers, by and large, made to order and they adopted a passive attitude to the market hoping that wholesalers would place orders when their travellers called. Today, except for its concentration in North Staffordshire and the wide range of its various product groups, the pottery industry is as interesting, or dull, depending upon which way one wants to look at it, as most other British industries. No longer is the industry a major source of pollution. It no longer condemns its work-force to long hours of arduous work for low pay. It conforms to type. The industry is operated by financial controllers, accountants and marketing men, profits are made and are reasonable. The bottle ovens have gone and, from the outside, the factories could be bakeries, or cigarette or carpet factories. Inside, however, they are distinctive; and this serves as a reminder that although things change, they can in fundamental ways remain the same. The message of this book is clear; the pottery industry in Britain is now, substantially, both modern and efficient. Year after year it will continue to make its not insubstantial contribution to the balance of payments.

Part VI

APPENDICES

APPENDIX I

BIBLIOGRAPHY

BOOKS

Technology

Chandler, Maurice, *Ceramics in the Modern World*, Aldus, London (1968).

Green A. T. and Stewart, G. H. (Eds.), *Ceramics: A Symposium*, British Ceramic Society, Stoke-on-Trent (1953). (Section on Pottery, pp. 212–415.)

History

Charleston, R. J., *World Ceramics*, Hamlyn, London (1968).

Hayden, Arthur, *Spode and His Successors: A History of the Pottery 1765–1865*, Cassell, London (1925).

Hellier, Bevis, *Master Potters of the Industrial Revolution: The Turners of Lane End*, Cory Adams & Mackay, London (1965).

McKendrick, Neil, 'Josiah Wedgwood and Cost Accounting in the Industrial Revolution', *The Economic History Review*, Second Series, Vol. 23, No. 1 (April 1970).

Nicholls, Robert, *Ten Generations of a Potting Family, Founded Upon 'William Adams, An Old English Potter' by William Turner*, Lund Humphries, London (Year of publication not stated).

Rhead, G. W. and F.A., *Staffordshire Pots and Potters*, Hutchinson & Co., London (1906).

Thomas, John, 'The Pottery Industry and the Industrial Revolution', *Economic History* (February 1937).

Thomas, John, *The Rise of the Staffordshire Potteries*, Adams & Dart, Bath (1971).

Warburton, W. H., *The History of Trade Union Organisation in the North Staffordshire Potteries*, Allen & Unwin, London, Ch. 2 (1931).

Copeland, R., 'Cheddleton Flint Mill and the History of Pottery Milling, 1726–1900', *North Staffordshire Journal of Field Studies*, Vol. 9 (1969).

Firms

Eyles, Desmond, *Royal Doulton 1815–1965: The Rise and Expansion of the Royal Doulton Potteries*, Hutchinson, London (1965).

Haggar, R. G., *The Masons of Lane Delph and the Origin of Mason's Patent Ironstone China*, Lund Humphries for G. L. Ashworth, London (1952).

Hatton, Joseph, *Twyfords: A Chapter in the History of Pottery*, J. S. Virtue & Co., London (Year of publication not stated).

Hollowood, Bernard, *The Story of J. & G. Meakin, 1851–1951*, Bemrose Publicity Co. (1951).

Kelly, Alison (in association with Josiah Wedgwood & Sons Ltd.), *The Story of Wedgwood*, Faber & Faber, London (1962).

Mackenzie, Compton, *The House of Coalport, 1750–1950*, Collins, London (1951).

Richards Tiles Ltd., *Richards, 1837–1953*, Stoke-on-Trent (1953).

Royal Academy of Arts, *200 Years of Spode*, London (1970).

Savage, George, *The Story of Royal Worcester and the Dyson Perrins Museum*, Pitkin Pictorials Ltd., London (1968).

The Industry

Board of Trade, *Pottery* (Working Party Report), H.M.S.O., London (1946).

Bryan, Arthur, 'Changes in the Ceramic Tableware Industry', *Royal Society for the Encouragement of Arts, Manufactures and Commerce* (November 1970).

Cooper, R. A., Hartley, K. and Harvey, C. R. M., 'The U.K. Pottery Industry', *Export Performance and the Pressure of Demand: A Study of Firms*, Allen & Unwin, London, Ch. 6 (1970).

Factory Inspectorate, *Industrial Health: A Survey of the Pottery Industry in Stoke-on-Trent*, Ministry of Labour and National Service: H.M.S.O., London (1959).

Gay, P. W. and Smyth, R. L., *The Household Demand for Tableware*, Department of Economics, University of Keele, Studies on the British Pottery Industry No. 3 (1970).

Gregory, Dennis and Smyth, R. L., *The Worker and the British Pottery Industry*, Department of Economics, University of Keele, Studies on the British Pottery Industry No. 4, (1971).

Jerrett, S. H., 'The Emergence of an Industry', *National Provincial Bank Review* (November 1967).

Joint Standing Committee for the Pottery Industry, *Pattern for Progress*, H.M.S.O., London (1972).

Jones, Mervyn, *Potbank: A Social Enquiry into Life in the Potteries*, Secker & Warburg, London (1961).

Lilleker, Christine and Smyth, R. L., *Women in the Pottery Industry*, Department of Economics, University of Keele, Studies on the British Pottery Industry No. 5 (1971).

Macdiarmid, Hugh and Smyth, R. L., *Exports of Pottery from the United Kingdom 1948–1968*, Department of Economics, University of Keele, Studies on the British Pottery Industry No. 1 (1969).

Machin, D. J. and Smyth, R. L., *The British Pottery Industry 1935–1968*, Department of Economics, University of Keele, Studies on the British Pottery Industry No. 2 (1969).

National Board for Prices and Incomes, *Pay and Other Terms and Conditions of Employment in the Pottery Industry*, Report No. 149, H.M.S.O., London, Cmnd. 4411 (1970).

National Society of Pottery Workers, *Reconstruction in the Pottery Industry*, Co-operative Printing Society, Manchester (1945).

Pottery Gazette and Glass Trade Review, *Pottery and Glass Manufacture and Selling: A Book for the Retailer and Student*, (Articles which first appeared in the Pottery Gazette), Scott, Greenwood & Son Ltd., London.

Restrictive Practices Court, *Glazed and Floor Tile Home Trade Association's Agreement, 1961*, No. 13 (E. & W.), 21 October 1963–17 January 1964.

Smyth, R. L., 'Theories of Competition and the British Pottery Industry', *Scottish Journal of Political Economy*, Vol. 18, No. 1 (February 1971).

Smyth, R. L., Irvine, V. and Gay, P. W., 'An Economic Survey of the British Domestic Pottery Industry', *North Staffordshire Journal of Field Studies*, Vol. 7 (1967).

Standing Committee on the Investigation of Prices, Sub-Committee, *Report on Pottery*, H.M.S.O., London, Cmd. 1360 (1921).

Tariff Commissions Report, Vol. 5, *The Pottery Industries, London*, P. S. King for the Tariff Commission (1907).

Williams, B. R., 'The Pottery Industry', in *The Structure of British Industry: A Symposium*, Duncan Burn (Ed.), Cambridge University Press, Vol. 2 (1958).

SOURCES OF BASIC STATISTICS

Pottery constitutes minimum list heading 462 of the *Standard Industrial Classification*, which is a part of Order No. XVI (bricks, pottery, glass, cement, etc.). As a result, for statistical purposes the industry is sometimes treated separately and on other occasions amalgamated with glass. There are three main sources of official statistics relating to the industry. The Department of Employment publishes in its *Gazette* data on hours of work, earnings, employment, etc. in which the classification is usually at M.L.H. level. This Department is also responsible for the *Retail Price Index* and the *Family Expenditure Survey*, in both of which pottery is amalgamated with glass.

The Department of Trade and Industry (formerly Board of Trade) is responsible for most aspects of production. The *Census of Production* contains a separate section on pottery, as does the quarterly *Business Monitor*. In the monthly *Index of Industrial Production* pottery is once again amalgamated with glass, but an annual figure is published for pottery separately. The Department of Trade and Industry also publishes in *Trade and Industry* a Wholesale Price Index in which Domestic China and Earthenware is a separate product.

Her Majesty's Customs and Excise identify several types of pottery separately in the *Overseas Trade Statistics*. Details of trade in pottery between other countries are available from the *United Nations Yearbook of International Trade*.

Many of the series mentioned above are re-printed in the *Annual Abstract of Statistics*, and those on output of foreign trade also appear in *Ceramics* and *Tableware International*.

PERIODICALS

Ceramics, Official Journal of the British Pottery Managers' Association, Stoke-on-Trent. Monthly.

Current Bibliography of Published Material Relating to North Staffordshire and South Cheshire, Stoke-on-Trent Libraries. Quarterly since 1964.

International Ceramic Industries Manual 1972, 2nd edition, Turret Press Ltd., London.

Tableware International, formerly Pottery Gazette and Glass Trade Review. Monthly since 1877.

Tableware and Pottery Gazette *Reference Book, 1972*. Tableware International. Annually.

APPENDIX 2

STATISTICAL SOURCES AND TABLES

This appendix contains notes on the sources of data and methods of construction of selected charts in the text, also some of the main statistical series.

Figure 1.1 The pottery industry 1920–1969: Index numbers of output and employment

Employment 1920–1938; A. L. Chapman and Rose Knight, *Wages and Salaries in the United Kingdom, 1920–1938* (Cambridge University Press 1953). Employment 1939–1969, *Annual Abstract of Statistics* (H.M.S.O., London). The employed are defined as the insured labour force less the unemployed.

The index of output was derived from gross output figures in the *Census of Production* (H.M.S.O., London). The value figures were deflated by a price index constructed as follows: 1920–1938, Richard Stone and D. A. Rowe: *The Measurement of Consumers' Expenditure and Behaviour in the United Kingdom, 1920–1938* (Cambridge University Press); 1939–1948, no price index of pottery available. Instead, the movements in the Wholesale Price Index of Manufactured Goods were used. The British Economy Key Statistics 1900–1970, London and Cambridge Economic Service (Times Newspapers Ltd.); 1949–1969, China and Earthenware Wholesale Prices, *Annual Abstract of Statistics* (H.M.S.O., London).

Note: To construct the series, retail and wholesale price indices and an index of prices of all manufactured goods were used to reflect movements in the factory prices of all pottery output. The series 1920–1967 is presented in Appendix B, D. J. Machin and R. L. Smyth, *The British Pottery Industry 1935–1968* (Department of Economics, University of Keele, 1969).

Figure 1.2   The pottery industry 1948–1969: Index numbers of output and employment

Year	(1) Numbers employed ('000s)	(2) Employment index (1948 = 100)	(3) Output
1948	75·5	100	100·0
1949	77·5	103	104·8
1950	80·5	107	109·5
1951	80·0	107	113·8
1952	78·9	105	108·2
1953	77·8	103	107·2
1954	78·4	104	108·8
1955	78·2	104	109·2
1956	75·7	100	98·4
1957	68·9	91	97·7
1958	68·0	90	99·6
1959	66·1	88	105·5
1960	67·1	89	108·8
1961	68·2	90	107·8
1962	68·9	93	108·8
1963	64·4	85	109·8
1964	66·0	87	123·2
1965	65·3	86	123·1
1966	65·0	86	126·0
1967	62·7	83	122·6
1968	60·0	79	130·1
1969	59·7	79	139·0

Sources: Col. 1, *Annual Abstract of Statistics*, H.M.S.O., London (employed in the insured labour force less unemployed); Col. 3, *Index of Industrial Production* (Pottery), *Annual Abstract of Statistics*, H.M.S.O., London.

Figure 1.3 The pottery industry 1948–1969: Index numbers of production

Year	(1) Pottery	(2) Value of output ($£'000$)	(3) Index of value	(4) Wholesale price index	(5) Output at 1948 prices (3) ÷ (4)
			Domestic pottery		
1948	100·0	21 680	100·0	100·0	100·0
1949	104·8	22 895	105·6	101·3	104·2
1950	109·5	24 677	113·8	102·8	110·7
1951	113·8	22 225	102·5	114·7	(89·4)
1952	108·2	29 388	135·6	121·8	111·3
1953	107·2	30 581	141·1	121·8	115·8
1954	108·8	31 422	144·9	126·6	114·4
1955	109·2	31 867	147·0	133·3	110·2
1956	98·4	28 195	131·1	136·7	95·9
1957	97·7	28 977	133·7	138·3	96·6
1958	99·6	32 018	147·7	145·5	101·5
1959	105·5	32 128	148·2	145·7	101·7
1960	108·8	33 894	156·3	150·2	104·0
1961	107·8	34 444	158·9	161·1	98·6
1962	108·8	35 161	162·2	167·9	96·6
1963	109·8	35 897	165·6	172·2	96·1
1964	123·2	38 575	177·9	177·3	100·3
1965	123·1	41 477	191·3	186·5	102·5
1966	126·0	42 850	197·6	197·6	100·0
1967	122·6	44 869	207·0	205·2	100·8
1968	130·1	48 529	223·8	214·3	104·4
1969	139·0	55 110	254·2	221·4	114·8

Sources: Col. 1, *Index of Industrial Production* (Pottery); Col. 2, Manufacturers' Sales of Manufactured Goods (domestic pottery); Col. 5, Wholesale prices (domestic china and earthenware), *Annual Abstract of Statistics*, H.M.S.O., London.

Figure 1.4 Comparative growth of the pottery industry and all manufacturing industry

The gross output of pottery at current prices was obtained from the *Census of Production* for census years. The price index was calculated as for *Table 1.1* with retail prices, 1907–1919, taken from A. R. Prest, *Consumers' Expenditure in the United Kingdom 1900–1919* (Cambridge University Press). The series for all manufacturing output was obtained from 'Index Numbers of Production in Manufacturing Industry', *The British Economy Key Statistics 1900–1970*, published for the London and Cambridge Economic Service by Times Newspapers Ltd. The series is presented in Appendix B, D. J. Machin and R. L. Smyth, *The British Pottery Industry, 1935–1968* (University of Keele, 1969).

Figure 5.2   Comparative growths of sales of types of tableware

Year	Earthenware	China	Stoneware	Jet and Rockington
1947	12·5	4·1	·4	·5
1948	15·7	5·0	·4	·6
1949	16·6	5·4	·4	·4
1950	17·4	6·4	·5	·4
1951	20·2	7·8	·7	·5
1952	19·8	8·4	·7	·5
1953	20·9	8·5	·8	·4
1954	21·2	8·9	·9	·4
1955	21·4	9·0	1·0	·5
1956	19·1	7·7	1·0	·4
1957	19·5	7·9	1·0	·4
1958	21·8	8·6	1·1	·4
1959	21·3	9·2	1·2	·4
1960	22·2	9·9	1·4	·4
1961	22·2	10·3	1·5	·4
1962	22·6	10·6	1·6	·4
1963	23·4	10·6	1·6	·4
1964	24·6	11·6	1·9	·4
1965	25·9	13·0	2·2	·4
1966	26·2	14·0	2·4	·3
1967	27·5	14·4	2·6	·3
1968	29·2	15·9	3·2	·3
1969	32·7	18·6	3·5	·3
1970	36·8	20·9	4·1	—

Source: *Annual Abstract of Statistics*, H.M.S.O., London and *Business Monitor* (Pottery), p.6 (1971 and 1972).

Figure 7.3   Total consumers' expenditure and expenditure on pottery in the United Kingdom 1920–1968

Expenditure on pottery for each year, 1920–1938, has been estimated by Richard Stone and D. A. Rowe in volume 2 of *The Measurement of Consumers' Expenditure and Behaviour in the United Kingdom 1920–1938* (Cambridge University Press). From 1948–1968 expenditure was estimated by adjusting the value of manufacturers' sales of domestic pottery, the *Annual Abstract of Statistics* for exports and imports, and adding a distribution margine of 60%. Richard Stone and D. A. Rowe used a distribution margin of 90% and their figures were adjusted to conform to a margin of 60%. *The British Economy*, *Key Statistics 1900–1970*, *op. cit.* contains annual figures for consumers' expenditure.

Figure 8.1 Pottery exports from the United Kingdom 1949–1971 '(£'000)

Year	Tiles	Sanitary ware	Electrical ware*	Domestic ware	Total
1949	1 658	2 920	919	12 265	17 762
1950	1 833	2 614	1 022	14 274	19 743
1951	2 810	3 350	1 146	16 264	23 570
1952	2 310	3 182	1 248	15 136	21 876
1953	1 836	2 405	937	12 592	17 770
1954	2 257	2 738	968	14 266	20 229
1955	2 579	3 509	1 076	15 014	22 178
1956	2 624	4 092	1 343	13 737	21 796
1957	2 170	3 231	1 167	12 909	19 477
1958	2 292	2 860	1 300	14 216	20 668
1959	2 483	2 910	1 187	14 077	20 657
1960	2 724	3 162	1 420	14 132	21 438
1961	3 077	2 732	1 569	14 304	21 682
1962	3 555	2 397	1 399	13 760	21 111
1963	3 750	2 554	1 158	13 951	21 413
1964	4 196	2 697	1 505	15 233	23 631
1965	4 327	3 107	1 610	16 555	25 599
1966	4 300	2 768	1 519	18 001	26 588
1967	3 679	2 773	893	18 358	25 703
1968	5 369	3 333	1 292	20 556	30 609
1969	6 232	3 796	1 452	25 061	36 541
1970	5 369	4 044	1 906	28 723	40 042
1971	5 867	—	—	32 408	—

Sources: *Trade and Navigation Accounts, Overseas Trade Statistics of the United Kingdom* and British Pottery Manufacturer's Federation.

*Electrical ware figures not comparable before and after 1962.

*Figure 9.1 Production of glazed tiles, bricks and cement and housebuilding:
Index numbers of output 1948–1969*

The index of the volume of tiles produced was calculated from the output of glazed tiles in 1000 square yards published in the *Annual Abstract of Statistics* (H.M.S.O.). Output in 1959 was 10 927 000 square yards and in 1969, 19 530 000 square yards.

Figure 10.1 Employment in the pottery industry 1948–1968

Monthly employment figures in the pottery industry are published in the Department of Employment *Gazette*.

*Figure 10.3 Male manual workers in pottery and all manufacturing industries:
Index of average hourly earnings*

Data relating to earnings are published by the Department of Employment at three-monthly intervals in *Statistics on Incomes, Prices, Employment and Production* (H.M.S.O.).

APPENDIX 3

AMALGAMATIONS AND TAKEOVERS IN THE POTTERY INDUSTRY

DOMESTIC WARE

Parent company	Subsidiary companies	Year of acquisition	Remarks
Allied English Potteries Ltd. (A subsidiary of Spearshaft Industrial Group Ltd. Ultimate holding company— S. Pearson & Son Ltd.	Booths Ltd.	1948	Consolidation of mainly family businesses into new organisation
	Colclough Ltd.		
	Ridgeway & Adderley	1953	
	Swinnertons Ltd.	1959	
	Bleak Hill Pottery		
	Alcock, Lindley & Bloore	1952	
	Thos. C. Wild & Sons Ltd.	1964	Renamed Royal Albert Ltd. in 1971
	Paragon China	1960	
	Royal Crown Derby Porcelain Co. Ltd.	1964	
	Shelley China Ltd.	1967	Closed production unit
	Doulton & Co. Ltd.	1971	Royal Doulton Tableware Ltd. formed in 1972

Parent company	Subsidiary companies	Year of acquisition	Remarks
Jon Anton Ltd.	Crown Clarence	1970	CWSEarthenware factory. Production concentrated in original factory
Avon Art Ltd.	Electra Porcelain Co.	1965	Merger. Production ceased in 1968
Aynsley China Ltd. (Untll 1971, John Aynsley & Sons (Longton) Ltd.)	Denton China (Longton) Ltd.	1969	Acquired by Waterford Glass in 1970
Joseph Bourne & Son Ltd. (Denbyware Ltd.)	Lovatts Potteries Ltd. (now Langley Potterery Ltd.)	1959	Bid for John Aynsley & Sons (Longton) Ltd. in 1970 not accepted
James Broadhurst & Sons Ltd.	Sampson Bridgwood Ltd.	1964	
R. T. Buckley & Co. Ltd.	Trent Walk Pottery	1959	
Burgess & Leigh Ltd. (Earthenware)	John Lockett Ltd. (Hospital ware)	1961	
Carborundum Ltd	W. T. Copeland & Sons Ltd. (Spode Ltd.)	1966	Diversification by an American company
	Hammersley & Co. (Longton) Ltd. (bone china)	1970	
	Windsor Bone China Ltd.	1971	
Cauldon Bristol Potteries Ltd.	Cauldon Bristol Pottery	1962	New factory erected in Cornwall in 1970
	Pountney & Co. Ltd.	1962	Acquired by A. G. Richardson & Co. Ltd. in 1972
Celmac Group Ltd.	Heatherley Fine China Ltd.	1968	
Alfred Clough Ltd.	Cartwright & Edwards Ltd.		
	Barker Bros. Ltd.	1959	
	W. H. Grindley & Co. Ltd.	1960	

Parent company	Subsidiary companies	Year of acquisition	Remarks
Consolidated Brick and Pipe Investments (New Zealand)	Royal Grafton China	1971	Originslly A. B. Jones & Sons Ltd. acquired by Crown House Investments Ltd. in 1966.
Doulton & Co. Ltd.	Dunn Bennett & Co. Ltd.	1968	
	Staffordshire Tea Set Co.	1966	Closed before take-over by Doulton. (Doulton acquired by S. Pearson & Son Ltd. in 1971)
	Minton Ltd.	1968	
	Webb Corbett Ltd. (Fine glass)	1969	
	John Beswick Ltd.	1969	
Dudson Bros. Ltd.	J. E. Heath Ltd. (Hotel ware)	1950	
	Grindley Hotel Ware Co. Ltd.	1952	
S. Fielding & Co. Ltd.	Shorter & Son Ltd. Baifield Productions Ltd.	1964	
R. Glew & Co. Ltd. (Subsidiary of Robin Wools of Bradford)	Jackson & Gosling Ltd. Grosvenor China Ltd.	1966	Closed 1969 ,, ,,
Govancroft Potteries Ltd. (Glasgow)	Manlan Pottery Ltd.	1963	
Great Universal Stores Ltd.	Barratt's of Staffordshire Ltd.	1948	
	Furnivals (1913) Ltd.	1967	Closed in 1969
Howard Pottery Co. Ltd.	Gibson & Sons Ltd.	1949	
	F. N. Lawrence (London) Ltd.	1956	
	Grimwades Ltd.	1963	
	Sudlow Ltd.		Closed down
	Norfolk Pottery		
Interpace Corporation (U.S.A.)	Myott, Son & Co. Ltd. (Earthenware)	1969	
Longton New Art Pottery	C. Amison & Co.	1963	Moulds only bought

Parent company	Subsidiary companies	Year of acquisition	Remarks
J. & G. Meakin Ltd. W. R. Midwinter Ltd.	Merger with Midwinter Merger with Meakin A. J. Wilkinson Ltd. Newport Pottery Co. Ltd.	1968 1964	Meakin & Midwinter (Holdings) Ltd. acquired by Wedgwood Ltd. 1970
Pearson & Co. (Chesterfield) Ltd.	Abbeydale New Bone China Co. Ltd. Ault Potteries Ltd. James Pearson Ltd. Price, Powell & Co. Ltd.		
S. Pearson & Son Ltd.	Doulton & Co. Ltd.	1971	Allied English Potteries Ltd. are also owned by Pearson
Pfaltzgraff Co. Division of Susquehanna Broadcasting Co.	Simpsons (Potters) Ltd.	1971	
Thos. Poole & Gladstone China Ltd.	Salisbury China Co. Ltd. British Anchor Pottery Co.	1961 1970	Controlled by the Gailey Group and trading as Hostess Tableware
Portmerion Potteries Ltd.	A. E. Gray & Co. Ltd. Kirkhams Ltd.	1961 1961	
Prestige Group Ltd.	Old Hall Tableware Ltd. (J. & J. Wiggin) Bridge Crystal Glass Co.	1970	
Qualcast Ltd.	Empire Porcelain Co.	1958	Pottery closed in 1967 after the acquisition of Qualcast Ltd. by Birmid Industries Ltd.
A. G. Richardson & Co. Ltd.	Pountney & Co. Ltd.	1972	Modern factory at Redruth acquired by a Stoke-on-Trent firm
Royal Worcester Ltd.	Palissy Pottery Ltd.	1958	Bid for Royal Worcester by Morgan Crucible rejected in 1971

Parent company	Subsidiary companies	Year of acquisition	Remarks
Semart Importing Co. (1966–Automatic Retailers of America)	Enoch Wedgwood (Tunstall) Ltd.	1964	
	Crown Staffordshire China Co. Ltd.		Acquired by Wedgwood in 1973
Sylvan Pottery Ltd.	Bournemouth Pottery Ltd.	1952	
	H. Tams Ltd.		
	H. Tams Ltd.	1949	
Staffordshire Potteries (Holdings) Ltd.	Collingwood China Ltd.	1955	All moved to one site at Meir Airport, Stoke-on-Trent. (Offer was made for Wood & Sons (Holdings) Ltd., in 1970, but it was not accepted)
	Chinaware Ltd.		
	Thos. Cone Ltd.		
	The Conway Pottery Ltd.		
	Keele Street Pottery Ltd.		
	Lawton Pottery Co. Ltd.		
	Paramount Pottery Co. Ltd.		
	Piccadilly Pottery Ltd.		
	Staffordshire Potteries Ltd.		
	Staffordshire Tableware Ltd.		
	Winterton Pottery (Longton) Ltd.		
John Tams Ltd.	Blyth Pottery	1939	
Taylor & Kent Ltd.	Rosina China Co. Ltd.		
	Ford & Sons (Crown Ford) Ltd.		Closed 1965
Thomas Tilling	Poole Pottery Ltd.	1971	Tilling acquired Pilkington Tiles Ltd. its subsidiary is Carter & Co. Ltd. and their subsidiary is Poole Pottery Ltd.
Victoria Porcelain (Fenton) Ltd.	Trentham Bone China Ltd.	1957	Ceased trading in 1960
Waterford Glass	Aynsley China Ltd.	1970	Spode and Denby also made bids for Aynsley
Wedgwood Ltd.	Royal Tuscan China	1966	Original company: Josiah Wedgwood & Sons Ltd., now part of the Wedgwood Group
	New Chelsea China	1961	
	Susie Cooper Ltd.	1966	
	Jason & Co. Ltd.	1950	
	William Adams & Son Ltd.	1966	
	E. Brain & Co. Ltd.	1967	

Parent company	Subsidiary companies	Year of acquisition	Remarks
Wedgwood Ltd. continued	Coalport Ltd.	1959	
	Johnson Bros. (Hanley) Ltd.	1968	
	Meakin & Midwinter (Holdings) Ltd.	1970	
	Kings Lynn Glass Ltd.	1969	
	Merseyside Jewellers Ltd.	1969	
	Crown Staffordshire China Co. Ltd.	1973	
	Mason's Ironstone China Ltd.	1973	
Arthur Wood & Son (Longport) Ltd.	Price & Kensington Pottery Ltd.		
	Carlton Ware Ltd.	1967	
Wood & Sons (Holdings) Ltd.	Wood & Sons Ltd.		
	Binsley Ltd.	1961	
	Corbett Goodwin Ltd.	1961	
	Middleport Mills Ltd.	1961	
	Ellgrave Pottery Co. Ltd.	1966	
	British Hotelware Supplies Heatmaster Ltd.	1956	
	H. J. Wood Ltd.	1956	

SANITARY WARE

Parent company	Subsidiary companies	Year of acquisition	Remarks
Armitage Ware Ltd.	Shanks Holdings Ltd.	1969	New company Armitage Shanks Ltd. (Name changed to Armitage Shanks Group Ltd., in 1972)
Doulton Sanitary Potteries Ltd.	Johnson & Slater Ltd.	1968	
Ideal-Standard Ltd.	John Steventon & Son Ltd.		Controlled by American Standard Inc.
Reed International Ltd.	Twyfords Holdings Ltd.	1971	(Bid by Glynwed rejected)

Parent company	Subsidiary companies	Year of acquisition	Remarks
Shanks Holdings Ltd.	J. & R. Howie Ltd. George Howson & Sons Ltd.	1964 1966	Now part of Armitage Shanks Ltd.
Wedgwood Ltd.	Johnson Bros. (Hanley) Ltd.	1968	Sanitary ware factory acquired along with table-ware manufac-turers

TILES

Thomas Tilling	Pilkington Tiles Ltd.	1971	
H. & R. Johnson Ltd.	Malkin Tiles Ltd. Richard Campbell Tiles Ltd.	1964/1968 1968	Name changed to H. & R. John-son–Richard Tiles Ltd. Richards Tiles Ltd. and Campbell Tiles Ltd. had merged in 1965

INSULATORS

Allied Insulators Ltd.	Bullers Ltd. Taylor, Tunnicliff & Co. Ltd.	1959	Two companies companies merged to become Allied Insulators Ltd.

APPENDIX 4

THE PRICING OF TABLEWARE

In 1939 two Oxford economists discovered, by interviewing business-men, that many favoured a mark-up on cost method of determining their prices.* The firms aimed to operate at near capacity output. To determine prices they would calculate their direct costs per unit of output—mainly labour and materials—and they would add to the average direct cost a mark-up to cover their overheads or fixed costs —salaries, rent, insurance, depreciation, etc. Finally another mark-up would be added to total average costs to cover profits. The prices so determined would be quoted to customers, and, if the firms enjoyed a degree of monopoly, the output would be sold at the cost-determined price. Hall and Hitch noted that adjustments would frequently be made to costings, production costs and products to arrive at prices which the market would accept. Some writers, subsequently, have tended to ignore the significance of the adjustments and have been inclined to claim that the prices of manufactured products are deter-mined entirely by costs and mark-ups.

P. W. S. Andrews undertook further empirical studies and, in *Manufacturing Business*, presented a similar theory of price determina-tion to that of Hall and Hitch. Andrews' 'Normal Cost Principle'

*R. L. Hall and C. J. Hitch, 'Price Theory and Business Behaviour', *Oxford Economic Papers*, No. 2 (1939).

operated as follows: 'The application of his costing rules and the resulting costing-margin will yield the business man what we shall call his costing price, and he will quote that price as a rule, never quoting above it in normal circumstances, and going below it only when the competition of others convinces him that he has made a mistake in the rightness of his costing rules.'*

R. H. Barback also published examples of firms which determined their prices by normal cost or cost-plus means.† Jack Downie accepted that businessmen prefer 'normal prices' in his book *The Competitive Process:* 'Firms have an idea of what constitutes a "normal" price or, more usually, a "normal" margin of profits over costs. That the notion of normal prices is a real one is evidenced by the fact that businessmen speak of "fancy" and "give-away" prices or of prices being at a level which can't last. Moreover, they refer to normal margins or mark-ups when questioned about how in fact they set their prices; to such an extent that there is now a school of "full-cost" price theorists among economists . . . the notion of normal price . . . provide(s) an anchor, which tends to confine price changes within not too wide a range.'‡

Normal cost theories of price are opposed to the theories of price determination which are to be found in most economics textbooks. Many economists do not accept cost-plus; instead, they state that firms aim to maximise profits and that their prices will reflect that objective. Their profits, therefore, will not be tied to costs and 'normal' profit margins. Furthermore in an industry dominated by competition, the market forces will determine prices and firms will be 'price-takers' rather than 'price-makers'.

In this appendix two statements on prices by pottery managers are presented. Both claim that the pricing of pottery tends to be a more complicated process than the 'normal cost principle' suggests. However, in addition, costings are presented from one fairly large pottery firm which has set its prices by a full-cost formula over a number of years. It would seem that many other pottery firms, particularly small ones, operate by means of similar procedures. The appendix concludes with the proposition that pricing can only be properly understood when it is viewed as one element in the marketing

*P. W. S. Andrews, *Manufacturing Business*, Macmillan & Co. Ltd., London pp. 158–159 (1949).

†R. H. Barback, *The Pricing of Manufactures*, Macmillan & Co. Ltd., London (1964).

‡Jack Downie, *The Competitive Process*, Gerald Duckworth & Co. Ltd., London, p. 112 (1958).

policies of firms. We believe that no matter what firms do, they *should* price not according to costs and desired profit margins, but according to the needs of the market. Firms should look outward—to the market—and not inward—at costings. Prices should be in accord with what customers are willing to pay, which may or may not agree with the results of normal cost formulas.

In 1969 R. A. Cooper, K. Hartley and C. R. M. Harvey asked pottery manufacturers how they determined prices.* They found that almost all the firms they interviewed raised their prices when costs increased rather than when market conditions changed. Their findings may be regarded as evidence that pottery manufacturers aim to earn reasonable profits rather than to maximise their profits.

COST–PLUS PRICING BY A TABLEWARE MANUFACTURER

The following details of a rigid cost-plus system were supplied by a tableware manufacturer who operates a fairly large firm. The direct cost is calculated for each article made. The margin for overheads is based on the historical relationship between total overheads and total direct costs. Having ensured that all costs will be covered, a further margin for profit is added. It is based on a profit requirement of an 18% return on capital employed and an historical capital–output ratio. From the point of view of the manufacturer, the pricing system provides a guarantee that every item produced contributes something towards both overheads and profits. The management concentrates on improving production techniques so as to keep costs as low as possible; this enables the firm to quote prices which are both profitable and highly competitive. After looking at how the firm calculates the cost of producing a plate we will note some of the limitations of the pricing system.

The firm records costs on a departmental basis, broken down into wages, materials, fuels, etc. For production departments a number of ratios are calculated, the 'on-cost' being the principal one for costing purposes.

$$\text{Percentage on-cost} = \frac{(\text{total costs}) - (\text{piecework payments})}{\text{piecework payments}}$$

*R. A. Cooper, *et al.*, *Export Performance and the Pressure of Demand: A Study of Firms*, Allen & Unwin Ltd., London, p. 108–109 (1971). For further details see page 162.

The on-cost represents the addition which must be made to the piece-work payments made in a department for an article to obtain an estimate of the total cost. For departments where payment is not on a piecework basis, total direct costs are related to some other variable which reflects the level of output; for example, for kilns, the total cost is related to the space occupied. The following is how the firm calculated the cost of producing a 6 inch white plate:

Clay: 4·66 new pence per dozen. The total direct cost of operating the clay preparation department and the weight of clay produced for each period is calculated. The cost per unit of clay is calculated. The weight of clay required to make one dozen plates is measured and the clay cost per dozen calculated.

Making: 4·82 new pence per dozen. Making operatives are paid on a piece-work basis. The figure is obtained by applying the on-cost for the making department to the piece-rates for the article.

Clay loss: 0·24 new pence per dozen. A 2·5% wastage rate on the cost already incurred.

Biscuit fire: 1·70 new pence per dozen. The total cost of operating the biscuit kiln and the adjacent warehouse is allocated to items on the basis of cubic capacity.

Biscuit loss: 1·14 new pence per dozen. The loss in the biscuit firing process is usually 10%.

Glost fire: 11·82 new pence per dozen. This is the cost of glazing the ware, the second firing and subsequent sorting in the warehouse.

Glost loss: 4·27 new pence per dozen. Losses usually 17·5%.

Total direct cost: 28·65 new pence per dozen.

Overheads: 11·46 new pence per dozen. 40% of direct cost.

Total cost: 40·11 new pence per dozen.

Profit: 10·03 new pence per dozen. This is 25% margin on cost. The firm requires an 18% return on capital. Given its capital/output ratio, this is equivalent to 20% on sales or 25% on cost.

Total: 50·14 new pence per dozen.

The cost, including profit, of an undecorated plate was 50·14 new pence per dozen. Decorated plates contain items for piece-rate payments and on-costs and decorating materials before the margins for overheads and profits are added. Customers who buy in large quantities expect trade discounts; 50·14 is the price customers would be expected to pay: additions are therefore added to this price so that 50·14 new pence would appear on invoices after trade discounts had been deducted. The customers who could not claim a trade discount paid more than the cost-plus price, their extra payments being regarded by the firm as a bonus.

In order that the mark-up on cost of 25% should result in the desired end of an 18% return on capital it is, of course, essential that planned levels of full-capacity production should be achieved. In practice, as the firm's profit record reveals, under-capacity working resulted in lower than expected profits and over-capacity working produced the opposite effect.* In addition, when the flow of production was interrupted by plant break-downs, new plant being run-in or labour shortages, then the extra costs were not passed-on to customers and the profits of the firm suffered. The crucial element in the situation is the level of costs and, as one would expect, the ability of the managers is probably more important than the particular pricing formula adopted. We suspect that, on occasions, the firm would ignore its self-imposed policy. A bit extra would be added to new and particularly attractive patterns and extra discounts or price reductions would be offered on well-established patterns. These aspects of pricing are given greater emphasis in the following two statements on pricing made by pottery managers.

Memorandum by the Chief Accountant of a Pottery Firm (1952)
In 1952 Donald S. Edwards published an article on 'The Pricing of Manufactured Products'.† The article concluded as follows: 'In discussion so many businessmen, and especially accountants, repeat the conventions described by the costing text-books, little realising how important it is that economists should be told about the adjustments that they are constantly making, or being forced to make, in order to meet the market valuation.' In the article some extracts from a memorandum proposed by the chief accountant of a pottery firm discussing the problem of price fixing in his industry were presented. The following are two of the extracts:

Price fixing is not a scientific operation, nor is it the result of a well defined mathematical formula. A mathematical formula or a cost estimate may be employed for arriving at a selling price as a basis of discussion, but it is my experience that costing, in relation to price fixing, is almost invariably a form of what psychiatrists

*When order books are full, then long-runs can be given priority over short ones and items which are relatively easy to fire can be substituted for difficult ones. The result is that the output per cubic foot of kiln space rises substantially; when orders are difficult to obtain the opposite tends to occur.

†D. S. Edwards, 'The Pricing of Manufactured Products', *Economica*, New Series, Vol. XIX, No. 75 (August 1952).

would call 'rationalisation'. That is to say, that the manufacturer has a 'hunch' as to the price at which his article can be sold, and makes use of 'costing' or 'estimating' to justify that price.

The 'hunches' of the manufacturer are not, however, to be disregarded. Although he may be unaware of the processes of thought by which he has arrived at that 'hunch' it is in fact the distilled essence of his knowledge and experience, and as such is of very much greater value for this purpose than all the involved calculations of the cost accountant. That is not to say that the accountant has no part at all to play in price fixing, but that he is by no means the final arbiter in the fixing of selling prices.

Where there is a range of articles offered for sale, there is always a scale of prices. In this industry the origin of the scale is usually lost in antiquity, and it was probably originally conceived very much on a 'hit and miss' principle. But over the years the scales have been adjusted from time to time and have become accepted both by the trade and the public as a true relationship between the various articles in the range. So that at today's date variations in price are, almost without exception, a bulk variation of the scale of prices and not variations within the scale. This acceptance of the existing relationship between articles in the scale does not necessarily mean that all the articles in that scale are equally profitable—far from it. Custom, and indeed commercial prudence, will demand a lower margin of profit on popular articles than on the speciality articles.

The following extracts are taken from a memorandum on 'Price Policy' written at our request. The firm specialises in ornamental ware.

Pricing is an art and not a science. Price, to those practical men who must set prices, is the result of an attempt to balance factors to which no precise weight can be attached. In particular he cannot precisely tell what competitors' reactions will be. Will competitors follow his price increase or decrease? How quickly will they react? Will they *all* follow him? Which ones might lag behind a price increase hoping to pick up volume at the old price? Should he lower his price to attract business? Which competitor is likely to beat him to the move? Will his price change be met exactly?

While pricing must be related to costs, we must not over-emphasise the place of costs in pricing. Costs are not always absolute truths, e.g. most manufacturers sell a line of products and not just one. It is therefore necessary to allocate many overhead costs to products.

The allocation bases are matters of judgement. Cost data represent actual *past* operations. All this does not mean to say that costs are not necessary and useful. Prices are set by what buyers are willing to pay. What they are willing to pay is influenced by their need and the availability, actual or potential, of competing or substitute products. Costs determine whether these are profitable to the seller. Prices, costs and volume are interacting forces. Volume is both a result of the other two factors and a determinant of them. Costs tend to establish price floors; demand tends to establish price ceilings.

The problem of price setting can be illustrated readily in the case of a manufacturer with a new, differentiated, product. Since the product is differentiated the manufacturer knows he need not price it identically with competitors' products. Since it is new he is not bound immediately to give consideration to the prices of competing substitute products, although these prices must ultimately be taken into account.

A useful initial first step is to determine who the buyers are and how much they might be willing to pay for the product. Next step (and more difficult)—how much *will* they pay? As far as this question is concerned the manufacturer might do some speculation on his own—get opinions of distributors and dealers, get buyers' evaluations. The latter might well be biased. All this kind of speculation leads to *a range of possible prices*. The manufacturer can then deduct from this range of reasonable prices the percentage margins that must go to dealers, etc. *The resulting figures are the prices that the manufacturer can place on his product.*

Piece costings are essential guides for pricing in firms which produce a wide range of products from a given plant. However, this does not mean that they should be used to determine prices by applying profit margins to them in a rigid manner. In the text we have noted some firms which allow market considerations to enter into pricing decisions: J. & G. Meakin, Wedgwood, Denby, Doulton and Royal Worcester; and of course, there are many others. Price fixing involves judgements about market conditions, product design, packaging, advertising and public relations, and channels of distribution. Most important, costs must be kept under strict control so that whatever mark-ups are used result in prices which are competitive. All these aspects of the activities of managers are somewhat inadequately summarised in the phrase 'full-cost pricing'.

Memorandum by the Managing Director of a Pottery Firm (1970).

APPENDIX 5

AN ANALYSIS OF THE DEMAND FOR TABLEWARE

In Chapter 7 we presented some results obtained from the tableware survey which we carried out in 1967–1968.* In this appendix some of the difficulties of analysing the demand for pottery will be discussed, and a brief description given of the survey and method of analysis of the data. Finally, some results more detailed than those presented in Chapter 7 will be given.

Domestic pottery is in some respects an ambiguous product. It is clearly a durable good, but because of its ability to be purchased in small quantities methods used for analysing the purchasing decision in relation to other consumer durables are not appropriate for pottery. For example, it has been suggested† that the ownership of a major consumer durable such as a car is a response (either present or absent) to such stimuli as an individual's level of income and wealth. Such an approach would be meaningless for pottery, since the decision made is not whether to own pottery or not, but how much to own, and the use of this type of analysis would require different types of sets to be identified as separate consumer durables.

---

*Most of the results presented in *Tables Appendix 5.1 to 5.7* were originally published by P. W. Gay and R. L. Smyth in *The Household Demand for Table-ware*, Department of Economics, University of Keele (1970).

†T. Cramer, *The Ownership of Major Consumer Durables*, Cambridge University Press (1962).

In addition, pottery is frequently bought not for the purchaser's own use, but for gifts, and this aspect of demand will obviously depend on quite different factors from those influencing purchases for own use, which themselves will be partly for replacement and partly to expand the stock of tableware possessed. Furthermore, it is clear both from our survey and from casual observation that to measure the stock of pottery *in use* is very difficult, since many householders are peculiarly reluctant to discard any item until it is absolutely useless. A difficulty shared by pottery with other consumer durables is the irregularity with which purchases are made, so that if a household is a low purchaser for the year studied, it is not necessarily always going to be so. If one adds to this list the point, frequently made to us by manufacturers, that pottery is to some extent a fashion good, it can be seen that any comprehensive explanation of demand and ownership behaviour would be extremely complex. What we have attempted is to isolate some of the main factors which appear to affect the level of purchases and the level of ownership.

THE TABLEWARE SURVEY

Early in 1968 a sample survey was carried out for us by Marplan Ltd. Full details of the questionnaire and method of analysis have been published elsewhere.* A brief description now follows.

The questionnaire, which was completed for 800 householders in urban areas throughout Great Britain, obtained data of two broad types, in addition to basic information on social class, family size, age and time since the establishment of the household. Firstly, a number of questions were asked relating to types of shop used for buying tableware, and shoppers' opinions of shop assistants, some of the results of which are given in Chapter 7. Secondly, householders were asked to give details of all purchases of tableware in the past year and to allow the interviewer to make a list of all items of tableware owned. For the purchases, details were obtained on price and reason for purchase, and for the inventory, etc. the age and condition of each item was recorded where possible.

Analysis of purchases and ownership data

There were two main aspects to this: firstly to obtain results relating

*P. W. Gay and R. L. Smyth, *The Household Demand for Tableware*, Department of Economics, University of Keele (1970).

to the sample as a whole, e.g. total expenditure; and secondly to analyse differences between households.

Purchases were recorded both as the number of items bought and the amount spent, while for the inventory only the number of items could be used as a measure of quantity. The validity of using the number of items as a measure of the stock of pottery was tested by examining the correlations, for purchases, between the number of items and the amount spent; the correlations are shown in *Table Appendix 5.9*.

Tables 7.1 and *7.5* and *Appendix Tables 5.1* and *5.2* summarise the properties of the sample as a whole. The average expenditure per household on purchases for all purposes, and including tableware made of materials other than pottery, was £3·74 for the year in question, which corresponds closely to the calendar year 1967.

Table 7.1 shows the breakdown of expenditure by material, both for total purchases and purchases for own use and gifts; the main point to note is the popularity of china and toughened glass for gifts.

Table Appendix 5.2 shows, for each material, the share of total expenditure which was for own use, and also the significance of replacement demand as a component of purchases for own use. Replacement demand is usefully considered in relation to *Table Appendix 5.2* and *Table 7.5* where the age distribution and condition respectively of the combined inventory are summarised.

The analysis of differences between households was carried out in two ways, firstly by an extension of the method used for producing the aggregate results, and secondly using regression analysis. The computer programs which analysed the survey data did so by examining the record for each household, accumulating totals for each material and then producing output which contained a list of each household and the totals relating to it. In addition, the household totals were accumulated to provide grand totals for the whole sample, which forms the basis of the tables so far discussed. This basic operation was carried out four times, each run collecting totals, both value and numbers of pieces, for each material. The four separate runs related to total purchases, own-use purchases, gifts purchases and ownership, and the output was written onto magnetic tape for further analysis. Since we had no preconceived ideas of what constituted heavy, light and medium ownership, etc. it was decided that the first analysis should be based on a ranking procedure, and accordingly for each of the totals produced, households were ranked from lowest to highest and the values taken by the households at the top of each decile were printed out. It should be pointed out at this stage that in much of the early analysis a distinction was made between items in sets and items which were odd pieces, and *Tables Appendix 5.3*, *5.4* and *5.5* show the results obtained for the main categories of demand and ownership.

The main point emerging from these is the unevenness of the distribution. For example, at least 640 of the 800 households (80%) spend nothing at all on items made of china (*Table Appendix 5.3*); and considering the grand total column of the same table, we find that 20% of the sample spent nothing at all and that the average household expenditure of £3·74 lies somewhere within the 8th decile. The irregularity with which purchases are made is obviously a strong contributory factor to this effect, which is even more marked when purchases for gifts are considered (*Table Appendix 5.4*), where we find that all purchases were concentrated among 30% of the households. Ownership (*Table Appendix 5.5*) is also unevenly distributed, although it is reassuring to know that no household owned less than 12 pieces. It would be interesting to discover what proportion of the 333 pieces recorded for the top household were in regular use.

The question which now poses itself is: 'To what extent are the variations between households in the level of purchases and ownership related to characteristics such as family size and social class, etc.?' The first attempt to answer this question was by means of a two-way analysis between, for each household, the decile of the distribution in which it was placed, and the attribute to which variations in the distribution might be related. *Table Appendix 5.6* shows the results obtained when this was done for ownership and those variables which seemed likely to affect it. Of the three, it can be seen that social class has more effect than the other two, and this was confirmed when the regression analyses were carried out (see below). It was also considered likely that level of ownership was related to level of purchases, and the two-way analysis of these two is shown in *Table Appendix 5.7*. The results indicate some tendency for high levels of purchases to be associated with high levels of ownership, as shown by the higher frequencies near the main diagonal of the table, but once again this relationship is more effectively demonstrated in the results of the regression analysis.

Regression Analysis

A small amount of further processing of the results was required in order to be able to use the household totals for purchases and ownership as input to a standard regression package. It should be pointed out at this stage that our survey did not measure all the variables we would ideally have used; in particular, we were not able to obtain data on householders' incomes. The best indicator of income was four-fold division by social class, but since this is not a variable measured on an equal-interview scale it could be incorporated in the equations only by the use of dummy variables. Furthermore, since, as has already been mentioned above, purchases are made irregularly and there are many influences on purchasing and ownership which cannot be measured, we did not expect to obtain a particularly good 'explanation' of variations in purchases and ownership, but nonetheless the exercise was useful in that it showed which of the hypothesised influences were significantly related and which were not.

The equations given include only those variables which have coefficients significant at the 5% level. *Table Appendix 5.9* shows the correlation matrix between some of the variables included in the regressions. The equations will now be examined individually for each relationship in turn.

1. *Total purchases**
$$X_1 = -38{\cdot}67 + 0{\cdot}63X_4 + 37{\cdot}40\,D_1 + 9{\cdot}50X_5$$
$$\phantom{X_1 = -38{\cdot}67 +} (0{\cdot}11) \quad (16{\cdot}24) \quad (3{\cdot}50) \quad R^2 = 0{\cdot}07$$

If total purchases are measured by number of pieces instead of value, then an equation with only two significant independent variables provides a better explanation of variations in purchases:

$$X_7 = -7{\cdot}07 + 0{\cdot}110X_4 + 3{\cdot}86X_5$$
$$\phantom{X_7 = -7{\cdot}07 +} (0{\cdot}014) \quad (0{\cdot}491) \quad R^2 = 0{\cdot}13$$

2. *Purchases for own use*
$$X_2 = -45{\cdot}33 + 0{\cdot}564X_4 + 10{\cdot}22X_5$$
$$\phantom{X_2 = -45{\cdot}33 +} (0{\cdot}098) \quad (3{\cdot}29) \quad R^2 = 0{\cdot}05$$

$$X_8 = -7{\cdot}45 + 0{\cdot}087X_4 + 3{\cdot}88X_5$$
$$\phantom{X_8 = -7{\cdot}45 +} (0{\cdot}013) \quad (0{\cdot}441) \quad R^2 = 0{\cdot}13$$

3. *Purchases for gifts*
None of the equations estimated had an R^2 of higher than 0·05, but of interest is the fact that of the variables with significant coefficients, family size no longer appeared, while D_1, the dummy associated with social class *AB*, was included, possibly indicating that the significance of D_1 in the total purchases equation was mainly associated with the gifts component.

*For all equations the figures in brackets are standard errors.

4. Ownership

$$X_4 = 83.58 + 59.15D_1 + 30.45D_2 + 13.27D_3 + 0.30X_6$$
$$\quad\quad (5.34) \quad\quad (4.53) \quad\quad (3.87) \quad\quad (0.13) \quad R^2 = 0.15$$

In many respects this is the most satisfactory of all the equations estimated. The constant value of 83·58 means that for a household in social class D or E, (working class, retired, etc.), which had been in existence for less than one year, 83·58 pieces of tableware would on the average be owned. The coefficient on D_1 shows by how much this would be increased if the household were classified as A or B (upper and middle class) and similarly the coefficients on D_2 and D_3 relate to lower middle class and skilled working class householders respectively. In addition, we could expect the stock of tableware to increase by 0·3 pieces for each year since the setting up of the household.

Table Appendix 5.1 EXPENDITURE ON EACH MATERIAL ANALYSED BY PURPOSE

Material	Own use expenditure as % of total	Replacement expenditure as % of total	as % of own use
China	70·2	13·3	19·0
Earthenware and stoneware	84·7	29·1	34·4
Plastic	67·2	11·7	17·4
Glass	67·1	16·7	24·8
All materials	72·9	20·2	27·7

Table Appendix 5.2 THE AGE STRUCTURE OF THE STOCK OF TABLEWARE (%)

Material	3 years	5 years	10 years
China	24·7	37·0	57·2
Earthenware	42·9	58·0	76·5
Stoneware	45·9	65·7	79·3
Plastic	66·9	84·1	95·9
Glass	48·7	64·7	85·0
All materials	39·7	54·0	72·9

The percentages show the share of the total stock of ware made of each material which is within the maximum age specified for each column.

Table Appendix 5.3 PURCHASES OF TABLEWARE, KITCHEN WARE (AND ORNAMENTS) BY VALUE BY PIECES, SETS AND MATERIALS (SHILLINGS)

Families ranked by value of their purchases	China		Earthenware		Stoneware		Plastic		Toughened glass		All materials		
	Pieces	Sets	Pieces	Sets	Pieces	Sets	Pieces	Sets	Pieces	Sets	Pieces	Sets	Grand Total
80th family	0	0	0	0	0	0	0	0	0	0	0	0	0
160th ,,	0	0	0	0	0	0	0	0	0	0	0	0	0
240th ,,	0	0	0	0	0	0	0	0	0	0	0	0	11
320th ,,	0	0	0	0	0	0	0	0	0	0	9	0	21
400th ,,	0	0	6	0	0	0	0	0	0	0	18	0	32
480th ,,	0	0	13	0	0	0	0	0	0	0	25	0	48
560th ,,	0	0	21	0	0	0	0	0	9	0	36	26	68
640th ,,	0	0	39	23	0	0	0	0	20	0	52	48	103
720th ,,	11	40	39	42	0	0	0	0	40	0	80	100	151
800th ,,	1 084	1 159	608	800	480	120	480	100	310	218	2 138	1 159	3 138

*All materials includes iron and steel which are not shown separately.

Table Appendix 5.4 PURCHASES OF TABLEWARE, KITCHEN WARE (AND ORNAMENTS) FOR GIFTS BY PIECES, SETS AND MATERIALS BY VALUE (SHILLINGS)

Families ranked by value of purchases	China		Earthenware		Stoneware		Plastic		Toughened glass		All materials		
	Pieces	Sets	Pieces	Sets	Pieces	Sets	Pieces	Sets	Pieces	Sets	Pieces	Sets	Grand Total
80th family	0	0	0	0	0	0	0	0	0	0	0	0	0
160th ,,	0	0	0	0	0	0	0	0	0	0	0	0	0
240th ,,	0	0	0	0	0	0	0	0	0	0	0	0	0
320th ,,	0	0	0	0	0	0	0	0	0	0	0	0	0
400th ,,	0	0	0	0	0	0	0	0	0	0	0	0	0
480th ,,	0	0	0	0	0	0	0	0	0	0	0	0	0
560th ,,	0	0	0	0	0	0	0	0	0	0	0	0	0
640th ,,	0	0	0	0	0	0	0	0	0	0	12	0	21
720th ,,	0	0	0	0	0	0	0	0	15	0	30	0	50
800th ,,	330	322	107	216	0	0	30	35	101	155	330	385	450

Table Appendix 5.5 OWNERSHIP OF TABLEWARE, KITCHEN WARE (AND ORNAMENTS) BY PIECES, SETS AND MATERIALS*

Families ranked by number of pieces owned	China		Earthenware		Stoneware		Plastic		Glass		All materials		
	Pieces	Sets	Pieces	Sets	Pieces	Sets	Pieces	Sets	Pieces	Sets	Pieces	Sets	Grand Total†
Family with least	0	0	0	0	0	0	0	0	0	0	0	0	12
80th family	0	0	3	0	0	0	0	0	1	0	14	18	55
160th ,,	0	0	8	7	0	0	0	0	2	0	21	32	69
240th ,,	0	0	12	18	0	0	0	0	3	0	26	40	81
320th ,,	0	18	17	21	0	0	0	0	4	0	32	51	89
400th ,,	2	21	22	26	0	0	0	0	5	0	38	59	100
480th ,,	3	21	27	36	0	0	1	0	7	0	43	70	112
560th ,,	6	36	33	43	0	0	2	0	9	0	49	81	122
640th ,,	9	42	41	51	0	0	4	0	11	0	60	96	143
720th ,,	16	60	52	66	1	0	7	0	15	15	72	124	168
Family with most (800th)	50	261	152	161	21	18	39	42	49	67	179	294	333

*Sets show total number of pieces in set or sets.
†Grand Total is total number of pieces in sets plus pieces not in sets.

Table Appendix 5.6 OWNERSHIP AND SOCIAL CLASS, LENGTH OF TIME FAMILY ESTABLISHED AND FAMILY SIZE (%)

Total number of pieces owned	Social class				Length of time family established (years)			Family size				
	AB	C_1	C_2	DE	0–14	15–29	30 and over	1	2	3	4	5 and over
1st decile	1·1	3·8	11·5	16·0	12·6	6·8	10·6	12·1	9·6	6·0	10·4	13·3
2nd „	2·1	5·6	12·5	13·0	12·0	6·8	8·8	18·2	7·2	8·0	10·8	11·1
3rd „	5·1	8·8	9·2	13·8	10·9	10·5	7·9	15·2	10·3	9·6	5·9	15·6
4th „	5·1	6·2	11·8	12·1	11·1	7·9	11·1	12·1	10·3	8·0	10·4	12·2
5th „	8·1	9·4	10·8	10·1	10·9	9·4	9·4	12·1	9·0	14·4	9·4	3·3
6th „	8·1	16·3	7·9	9·1	8·3	10·5	12·2	15·2	10·3	11·2	9·0	4·4
7th „	7·2	14·3	9·9	8·4	8·3	11·2	11·7	9·1	11·4	9·6	9·9	7·8
8th „	11·3	10·0	8·9	10·8	10·3	10·2	9·4	—	12·7	10·7	9·4	16·7
9th „	22·0	13·8	8·9	4·2	8·8	10·9	11·1	3·0	11·4	10·7	14·4	5·6
10th „	29·9	11·8	8·6	2·5	6·8	15·8	7·9	3·0	7·8	11·8	10·4	10·0
Total	100·0	100·0	100·0	100·0	100·0	100·0	100·0	100·0	100·0	100·0	100·0	100·0

Table Appendix 5.7 TOTAL PURCHASES IN ONE YEAR (1967), OF TABLEWARE, CHINA, EARTHENWARE, PLASTIC AND TOUGHENED GLASS MEASURED BY NUMBER OF PIECES (INCLUDING PIECES IN SETS) AND THEIR RELATIONSHIP TO TOTAL OWNERSHIP OF TABLEWARE SIMILARLY MEASURED*

Ranking of families by purchases	Ranking of families by total ownership										Total
	80	160	240	320	400	480	560	640	720	800	
80	11	12	12	15	11	11	3	4	1	0	80
160	3	12	7	4	6	15	8	10	13	2	80
240	14	7	9	1	7	7	11	7	9	8	80
320	6	4	9	11	15	4	8	8	9	6	80
400	7	9	5	9	7	10	9	13	7	4	80
480	12	8	8	11	6	4	4	6	11	10	80
560	12	4	8	5	8	11	8	8	6	9	79
640	5	6	6	11	11	9	9	8	8	7	80
720	6	11	5	6	3	7	11	6	8	17	80
800	4	7	11	7	6	2	9	10	8	16	80
Totals	80	80	80	80	80	80	80	80	80	79	

*Each entry in the table represents the number of households who are ranked according to the row heading and the column heading, e.g. the entry 12 in column 2 of row 1 indicates that 12 households were ranked between 81st and 160th for ownership and between 1st and 80th by purchases.

Table Appendix 5.8   VARIABLES USED IN THE REGRESSION ANALYSIS

Symbol	Variable		Mean	S.D.
X_1	Value of total purchases	Measured in shillings (5p)	67·66*	146·97
X_2	Value of own use purchases		52·27	136·95
X_3	Value of gifts purchased		15·39	39·77
X_4	Total number of pieces owned		107·53	48·10
X_5	Number of persons in household		3·61	1·44
X_6	Number of years since establishment of household		18·40	11·88
X_7	Total number of pieces purchased		18·73	21·36
X_8	Total number of pieces purchased for own use		15·94	19·22
X_9	Total number of pieces purchased for gifts		2·79	8·33
D_1		= 1 if class A,B, otherwise = 0		
D_2	Social Class Dummies = 1 if class C_1, otherwise = 0			
D_3		= 1 if class C_2, otherwise = 0		

*This value is lower than that quoted earlier (£3.74). Some householders were unable to specify the price paid for some of their purchases; in the aggregate analysis a record was kept of the number of unpriced items and the total expenditure figure adjusted upwards on the assumption that the average price per piece of the items not priced was the same as that for items for which prices were recorded.

Table Appendix 5.9 CORRELATION MATRIX FOR SELECTED VARIABLES

	X_1	X_2	X_3	X_4	X_5	X_6	X_7	X_8	X_9
X_1	1·000								
X_2	0·9632	1·000							
X_3	0·3787	0·1160	1·000						
X_4	0·2310	0·1974	0·1740	1·000					
X_5	0·0899	0·1060	−0·0326	−0·0067	1·000				
X_6	−0·0426	−0·0644	0·0643	0·0719	−0·3395	1·000			
X_7	0·6009	0·5278	0·4032	0·2461	0·2584	−0·0487	1·000		
X_8	0·5560	0·5740	0·0784	0·2158	0·2889	−0·0979	0·9211	1·000	
X_9	0·2571	0·0283	0·8528	0·1330	−0·0042	0·1011	0·4380	0·0533	1·000

INDEX

Adams & Son Ltd., William, acquisition by Wedgwood, 44, 65
Advertising, 145–146
Allied English Potteries Ltd:
  acquisition of Royal Crown Derby, 87
  formation of, 45–46, 58, 74
  Group Administrative Centre, 78
  individual production divisions described, 76–77
  low profit rate 1968–1971, 80
  performance in 1960s, 74–75
  reorganisation in 1969, 75–76
  transfer of control to Doulton Fine China, 78
Allied Insulators Ltd:
  dominance of electrical wave sector, 243
  formation of, 42
  performance 1961–1970, 194
Amalgamations:
  in pottery industry, 46–49
  in tile sector, 168
  see also for specific amalgamations, entries for individual firms and Appendix 3
Andrews, P. W. S., 266, 267n
Armitage Shanks Ltd., 42, 185
Armitage Ware Ltd., 186, see also Armitage Shanks Ltd.
Australia, as importer of pottery, 157
Automatic Retailers of America, see Semart Importing Company
Aynsley & Sons Ltd., John, 48

Bailey, R. J., 78
Barback, R. H., 267
Barker Brothers Ltd., 95
Barratts of Staffordshire Ltd., 48
Beswick Ltd., John, 45
Biltons (1912) Ltd., 245
Blauner, R., 213n
Body compositions, 20
Bone china, 81, 85, 86
Boote Ltd., T. & R., 168
Booths Ltd., 45
Boulton (Holdings) Ltd., William, 13
Bourne, Joseph, 15 see also Denbyware
Bowers, E. C., 92
British Ceramic Manufacturers' Federation, 51, 221
British Ceramic Research Association, 15, 52, 214
British Ceramic Tile Council Ltd., 51
British Electro-Ceramic Manufacturers' Association, 51
British Pottery Manufacturers' Federation, 88
  see also British Ceramic Manufacturers' Federation
British Pottery Promotion Service Ltd., 146
Bryan, Arthur, 36, 65
Bullers Ltd., 42 see also Allied Insulators Ltd.

Campbell Tile Co. Ltd., 168

Canada, as importer of pottery, 156
Capital intensity, 121, 232
Carborundum Ltd., 86, 244
Carlton Ware Ltd., 110
Carter Tiles Ltd., 168
Cartwright & Edwards Ltd., 95
Ceramic and Allied Trades Union, 15, 51, 221–222
Chapman, A. L., 254
Charleston, R. J., 58n
Clough, Alfred J., 92, 96, 97, 98
Clough Ltd., Alfred, 85, 95, 96, 97, 98
Clay:
    as raw material, 19–21
    mixing of different types, 23n
    producers of, 12, 13
    properties of, 19–20
Coalport Ltd., 49, 65
Colclough Ltd., 45
Competition:
    and elimination of high-cost firms, 39
    in pottery industry, 15–16, 36–52
    in sanitary ware sector, 183–184
    in tile sector, 167–168
    'workable', 242–243
Concentration:
    geographical, 14–15
    in pottery industry, 41–43
    in sanitary ware sector, 184–185
    in tile sector, 108
Cooper, R. A., 157n, 162n, 268
Cooper Ltd., Susie, 44
Cooper-Willis, Mrs. J., see Williams-Ellis, Susan
Copeland, W. T. & Sons Ltd., 47, 86 see also Spode Ltd.
Copeland, William, 86
Corner, D. C., 40n
Cost-plus pricing, 267–270
Costs:
    allocation of shared costs, 126–129
    in relation to pricing policy, 129–130
    of substandard ware, 130–131
    overheads, 126–128
    reduction by means of standardisation, 128, 129
    share taken by imports, 125–126
    structure of, 119–125
    transport as factor favouring disposal of sanitary ware factories, 179
Cotton, Elijah, 111
Cotton Ltd., Elijah, 85, 111–112, 244
Council of Britain Ceramic Sanitaryware Manufacturers, 51, 182
Cramer, T., 273n
Crapper, Thomas, 179

Crazing, 20, 31
Crown Staffordshire China Co. Ltd., 44, 48, 66

Dartington Glass Ltd., 113
Decorating see Manufacturing processes
Demand:
    for pottery, stability of, 238–240
    for tableware, analysis of, 273–285
    see also Expenditure on tableware
Denbyware Ltd:
    design policy, 115
    expansion of stoneware sales, 115
    marketing policy, 117
    shops within shops, 63
    subsidiary companies, 114
Department stores:
    and shops within shops, 63, 134
    as outlet for pottery, 61, 133–134
Devaluation:
    of dollar in 1971, 163
    of pound in 1967, 8, 148, 161
Die-pressing technique, 198
Distribution of pottery:
    changes in methods, 132
    decline of wholesaling, 132
    department stores and supermarkets, 133–134
    margins in, 133
    stores within stores, 134–135
Diversification:
    as component of growth of firms, 40
    of pottery industry, 3
    within groups in pottery industry, 244
Doulton & Co. Ltd:
    abandonment of earthenware bodies, 72
    acquisition by Pearson, 46, 69
    acquisitions of:
        Johnson & Slater Ltd., 42, 185–186
        Whieldon Sanitary Potteries, 179
    as producers of insulators, 42
    development of ETC, 71–72
    diversification within, 4
    establishment in Burslem, 69–70
    expansion in 1960s, 72
    formation and growth, 44–45
    geographical dispersion, 70
    importance of research, 70–71
    in relation to competition in domestic ware, 243
    increasing specialisation, 244
    management quality, 100
    rapid changes since 1950, 87
    reorganisation after acquisition by

Doulton & Co. Ltd. *continued*
   Pearson, 69
   shops within shops, 63
   structure and performance of table-
      ware division, 73
Doulton Industrial Products Ltd.,
   195–197
Doulton Insulators Ltd., 191
Doulton, Henry, 56, 179
Doulton, John, 55
Downie, Jack, 167n, 267
Dunn Bennett & Co. Ltd., 45
Dunn, W. B., 81

Earnings:
   level of, 208, 209–211
   structure of, 207
Edwards, R. S., 270
Electrical ware:
   dependence on exports, 193
   manufacturing methods, 190–191
   post-war boom, 192
   recession and spare capacity, 192
Empire Porcelain Company, 48
Employment:
   by sector, 3, 4, 5
   expected further decline in, 205, 236
   opportunities in Potteries, 14
   per factory in various years, 11
   total for industry, 203–205
English Bone China Manufacturers'
   Association, 51
English China Clays Ltd., 12–13
English Glass Company Ltd., 200
Entry into the industry, 34, 50, 242
Europe, as market for British pottery,
   162–163
Evans, W. G., 89
Excess capacity in electrical ware sectors
   192
Expenditure on tableware:
   changing priorities for, 142n
   dependence on household income, 138
   distribution among materials, 136
   increase in 1960s due to price increases,
    137
   irregularity of, 142
   low level of, 135, 142, 144
   related to ownership, 144
   related to total consumers' expendi-
    ture, 141
   suggestions for increasing, 144–145
   value of, 136
   see also Demand for tableware

Export of pottery:
   analysis of sources and destinations,
    149–151
   dependence of British industry on,
    10–11, 152
   description of American market,
    159–161
   effects of dollar crisis, 163
   effects of E.E.C. membership, 162–163
   export performance and the pressure of
    demand, 161–162
   export performance compared with
    other industries, 152
   export performance of industry des-
    cribed, 148–149
   long-run prospects for, 241–242
   methods of distribution, 157–159
   of electrical ware, 193
   origins of, 152
   poor performance since 1955, 153–156
External economies, 12–13
Eyles, Desmond, 56n, 179n

Factory size, 11, 12
Family businesses, 36–38
Faraday, Samuel Baylis, 89
Fireclay sanitary ware, 189–190
Firing:
   biscuit, 27–30
   glost, 31
   one piece, 30
   temperatures for electrical ware, 190
   temperatures for various bodies, 21
   see also Kilns
Firms, *see* individual entries
Fletcher, R., 102
Freeman, P. H., 105
Full-cost pricing, 266, 267, 272
Furnivals Ltd., 48

Galbraith, J. K., 49n
Gay, P. W., 136n, 142n, 239n, 273n,
   274n
General Earthenware (Home and Export)
   Manufacturers' Association, 51
Germany, West, as exporter of pottery,
   151–153, 157
Gifts, as reason for purchasing tableware,
   136–137
Glass tableware, 137
Glazed and Floor Tile Home Trade
   Association, 168
   approval of agreement, 169
   details of agreement, 169–170

Glazing, see Manufacturing processes
Goddard, J. S., 89
Godden, G. A., 89n
Govancroft Potteries Ltd., 245
Great Universal Stores Ltd., 48
Green Ltd., T. G., 85, *105–106*, 244
Gregory, D. L., 216n, 219n
Grindley Ltd., W. H., 95
Growth of firms, 39–40
Grundy, J. W. E., 100

Haggar, R. G., 89n
Hall, R. L., 266n
Hartley, K., 158n, 162n, 268
Harvey, C. R. M., 158n, 162n, 268
Hatton, Joseph, 187n
Hepworth Iron Ltd., 44
Hill, A. B., 212n
Hitch, C. J., 266n
Hodson, S. S., 111
Hollowood, Bernard, 101
Hornsea Pottery Co. Ltd., 245
Hostess Tableware, 100
Hotel ware, 135–136
Hunter, Alex, 242n

Ideal-Standard Ltd., 42, 185, 187
Imports of pottery:
    analysis of buyers and suppliers,
        149–151
    by Australia, 157
    by Canada, 156
    by U.S.A., 155
    see also Exports of pottery
Industrial Ceramics:
    described, 195
    Doulton Industrial Products Ltd.,
        195–197
    Royal Worcester Ltd., 197
    Wade Potteries Ltd., 197–200
    see also Electrical ware, Sanitary ware,
        Tiles
Industrial diseases, 21n, 214–215
Industrial relations 220–227, 240–241
    annual wage settlement, 222–224
    employers' and workers' organisa-
        tions, 220–222
Integration, vertical, 50
International Ceramics Ltd., 114

Jasper ware, 39
Jerrett, S. H., 52

Job satisfaction:
    among men, 213
    among women, 211–213
Johns & Co. Ltd., Edward, 186
    see also Armitage Ware Ltd.
Johnson, Derek H., 168, 169, 175
Johnson Brothers (Hanley) Ltd:
    acquisition by Wedgwood, 44, 65
    acquisition of J. & G. Meakin Ltd., 44
    acquisition of W. R. Midwinter Ltd.,
        44
    as producers of sanitary ware, 179
Johnson, H. & R. Ltd:
    acquisition of Malkin Tiles Ltd., 168
    merger with Richards Campbell Tiles
        Ltd., 41, 168
Johnson-Richards Tiles Ltd., H. & R.:
    dominant position of, 41, 243
    formation of, 168
    growth and performance, 176–177
    innovating activities, 175
Johnson & Slater Ltd., 42, 179, *185–186*
Johnson, William, 111n
Jones, A. Derek, 100
Jones, Mervyn, 15

Kent, William, (Porcelains) Ltd., 199
Kilns:
    biscuit, 27–29
    enamel, 31
    glost, 31
    hover, 30
    intermittent:
        bottle (coal-fired), 86
        gas-fired, 86
    top-hat, 30
    tunnel, 29–30, 38
Knight, Rose, 254
Kusmirek, A., 239n

Labour intensity:
    and labour productivity, 232
    in pottery manufacture, 119, 209
    in retailing, 134
    predicted decline in, 236
Labour force, limited bargaining power
    pre-1945, 37
Labour productivity, see Productivity,
    labour
Labour relations, development in tile
    sector, 176
    see also Industrial relations
Labour turnover:
    and attitudes among school leavers,
        219–220

Labour turnover: *continued*
  high level among young workers, 218
  high level of, 216
  pottery compared with other indus-
    tries, 217
  related to labour scarcity, 216
Lawley Group, 45
Lawleys Ltd., 77–78
Langley Pottery Ltd., 114
Lilleker, Christine, 211n
Location:
  of pottery industry, 14
  of sanitary ware sector, 179
Loftus, P. J., 120

McCann-Erickson, 60n
Machin, D. J., 138n, 254
McKendrick, N., 56n
Malkin Tiles (Burslem) Ltd., 168
Manufacturing processes:
  body preparation, 23–24
    dust for tile making, 24
    plastic clay, 23–24
    slip for casting, 24
  firing, 27–30
  for electrical ware, 190–191
  glazing and decorating, 31–33
  making, 25–27
    casting, 26
    dust pressing and die stamping,
      26–27, 171–172
    jiggering, 26
    jollying, 25
  packing, 33
Mason, Charles James, 88, 89
Mason, George Miles, 89
Mason, Miles, 89
Mason's Ironstone China Ltd., 66, 85,
  88–90
Meakin Ltd., J. & G.:
  delivery problems, 102, 103, 104
  merger with Midwinter, 44, 65, 101,
    105
  new management, 100, 101
  pricing policy, 104, 105
  simplification of product range, 103,
    129
  suppliers of middle range of market,
    85
Mergers, 243–244
  and decline of family firms, 38
  see also for specific mergers, entries for
    firms involved and Appendix 3
Meyer, F. V., 40n

Midwinter Ltd., W. R.:
  benefits from product rationalisation,
    129
  merger with J. & G. Meakin and
    acquisition by Wedgwood, 44,
    65, 101
Millard Norman Co., 114
Minton, Thomas, 88
Monopoly in insulator market, 42
Morley, Francis, 89
Myers, Lucien, 63n

Nash, J., 107
National Joint Council for the Pottery
  Industry, 51, 222–223
Normal Cost Principle, 266–268
North Staffordshire College of Tech-
  nology, 60n

Once-fired ware:
  tableware, 22 30, 92
  sanitary ware, 180
Ornamental Pottery Association, 51
Ornamental ware, 85
Output:
  contraction and fluctuation in 1920s, 5
  growth between 1907 and 1963, 9
  of pottery industry in 1963 and 1970, 4
  post-war boom, 6
Ovens, *see* Kilns

Packing, 33
Palissy Pottery Ltd., 81
Parker, J. E. S., 40n
Paragan Bone China Ltd. *see* Thomas C.
  Wild & Sons Ltd.
Pay, *see* Earnings
Pearson Group, 84, 244
Pearson, Miss J., 239n
Penrose, E. T., 40
Perrins, Charles W. D., 58
Pilkington Tiles Ltd:
  acquisition by Tilling, 178
  growth since 1960, 168, 177–178
  in relation to competition in tiles,
    41, 243
Plant Ltd., R. H. & S. L., 44
Plastic tableware, 137
Polycell Products Ltd., 178
Poole Pottery, 15, 63
Poole and Gladstone China Ltd.,
  Thomas, 85, 106–108, 245

Porcelain, hard:
  production, 81
  marketing, 81
Portmerion Potteries Ltd., 85, 112–113, 245
Poultney & Co. Ltd., 245
Prest, A. R., 256
Price and Kensington Potteries Ltd., 110
Pricing of pottery, 146–147
  tableware pricing methods, 266–272
    cost-plus, 267–270
    full-cost, 266, 267, 272
    normal cost principle, 266–268
  tiles:
    bulk discounts, 174
    pricing agreements, 169–170
Printing machine, Murray Curvex, 86
Productivity, labour:
  changes in, 6–8, 227–232
  comparisons with other industries, 229–232
  factors affecting, 227
Purchases of tableware, influences on, 145
  see also Expenditure on tableware

Qualcast Ltd., 48

Reed International Ltd., 42, 188
Restrictive Practices Court, 168n, 169
Richard Campbell Tiles Ltd., 41
Richards Tiles Ltd., 168
Richardson & Co. Ltd., A. G., 245
Ridgway, C. C., 212n
Ridgways Ltd., 46
Ridgway Hotelware, 77
Ridgway Potteries Ltd., 77
Ridgway Royal Adderley Floral, 77
Rowe, D. A., 254, 257
Royal Albert Ltd., 129
  see also Thomas C. Wild & Sons Ltd.
Royal Art Pottery, 95
Royal Crown Derby Porcelain Co. Ltd., 15, 46, 76–77, 87
Royal Worcester Ltd:
  as producer of industrial ceramics, 195, 200
  description of company, 80–83
  description of hard porcelain, 21
  diversification into electronics, 244
  in relation to competition in domestic ware, 243
  origins and growth of company, 58
  shops within shops, 63

Savage, George, 58n
Sanitary ware, 178–190
  competition, 183–184
  current structure of industry 184–190,
  early growth of industry, 178, 179
  factors affecting level of demand, 180–182
  fireclay, 189–190
  marketing methods, 182–183
  sources of demand, 179–180, 181–182
  technical progress in manufacturing, 180
Scale:
  diseconomies of, 39
  economies of:
    and decline of family firms, 38
    in pottery industry, 12
    in supply of body materials, 23
Scarratt, W., 118
'Seconds' see Sub-standard ware
Shanks Ltd:
  acquisition by Armitage, 185
  acquisitions in 1960s, 185
  development of battery method of production, 186
  location and specialisation, 179
  see also Armitage Shanks Ltd.
Shops within shops, 134–135
  Denby, 117
  Doulton, 63
  Royal Worcester, 63
  Wedgwood, 63
Semart Importing Company, 48, 244
Simpsons (Potters) Ltd., 245
Skill, 205–209
Slip:
  defined, 23n
  production of, 23, 24
Small firms, future prospects for, 244–245
Smith, Arnold, 177
Smyth, R. L., 136n, 138n, 142n, 211n, 216n, 219n, 239n, 254n, 256n, 273n, 274n
Specialisation, 34–35
Spode I, Josiah, 86
Spode II, Josiah, 86
Spode Ltd., 48, 58, 85–88
Staffordshire Potteries (Holdings) Ltd., 48, 90–95, 100
Staffordshire Potteries Hotel Ware Manufacturers' Association, 51
Standardisation:
  in sanitary ware, lack of, 184
  in tiles, 41, 169, 172–173

Stock of tableware:
  age composition of, 143
  condition of, 143
  related to purchases, 144
Stone, Richard, 254, 257
Stores within stores
  see Shops within shops
Structure of pottery industry, 3–17
Sub-contracting:
  among workers, 37
  in electrical engineering, 192
Sub-standard ware:
  tableware 130–131, 144
  tiles, 169–170
Supermarkets, 134
Swinnertons Ltd., 46

Takeovers see Amalgamations
Taylor Tunnicliff & Co., 42
  see also Allied Insulators Ltd.
Technology of pottery industry, 18–35
Tiles, 167–178
  as example of Downie's competitive
    process, 167–168
  details of pricing agreements, 169–170
  growth and performance of Johnson-
    Richards, 176–177
  growth of wholesaling function, 174
  manufacturing methods, 171–172
  mergers and takeovers, 168
  output related to building activity, 175
  recovery of Pilkington Tiles, 177–178
  reduction in number of firms, 168
  reductions in prices, 173–174
  Restrictive Practices Court decision,
    169
  standardisation of product, 169,
    172–173
  technical improvements, 172–173
Trade associations, 51
  see also individual entries
Twyfords Ltd., 42, 179, 187–189

U.S.A.:
  as importer of pottery, 155, 159–161
  effect of dollar crises on pottery
    imports, 163

Vitreous ware, 136

Wade, A. J. Ltd., 197
Wade, Col. Sir G. A., 199
Wade Heath & Co. Ltd., 197, 199
Wade Potteries Ltd:
  activities and performance, 198
  changes in product range, 197, 200
  die-pressing technique, 198
  establishment of factory in N. Ireland,
    199
  industrial ceramics, 195
  processing of alumina, 200
Wage determination, 224–227
Wall, Dr. John, 58
Watts, Blake, Bearne & Co. Ltd., 13
Webb Corbett Ltd., 45
Wedgwood (Tunstall) Ltd., Enoch, 48
Wedgwood I, Josiah, 39, 55, 56
Wedgwood V, Josiah, 58
Wedgwood VI, Josiah, 56
Wedgwood Ltd:
  designers, 60
  expansion of capacity at Barlaston, 66
  formation of, 44
  in relation to competition in domestic
    ware, 243
  management quality, 100
  marketing strategy, 61–63
  personnel department, 64
  promotion of product, 60–61
  takeover activities, 44, 65–66, 101
  Wedgwood Rooms, 63
Welwyn Electric Ltd., 197
Western Gold & Platinum Company,
  200
Wholesalers:
  in export markets, 157–158
  in home market, 132–133
  in sanitary ware distribution, 182–183
  increased use of for tiles, 174
Wild, Kenneth, 105
Wild & Sons Ltd., Thomas C., 46, 77,
  105
Wilde, R., 212n
Williams, B. R., 46n, 209
Williams-Ellis, Susan, 113
Wood, A. F., 109
Working conditions, 214–215
Wood & Son (Longport) Ltd., Arthur,
  85, 109–111, 245